T0338260

BOW TIES IN RISK MANAGEMENT

This book is one of a series of process safety guidelines and concept books publish ed by the Center for Chemical Process Safety (CCPS). Please go to www.wiley. com/go/ccps for a full list of titles in this series.

This concept book is issued jointly with the Energy Institute. In EI publications, concept books are termed Research Reports in its series of technical publications. EI publications can be found at http://publishing.energyinst.org/.

BOW TIES IN RISK MANAGEMENT

A Concept Book for Process Safety

CCPS in association with the Energy Institute

CENTER FOR CHEMICAL PROCESS SAFETY
OF THE
AMERICAN INSTITUTE OF CHEMICAL
ENGINEERS
New York, NY
and
ENERGY INSTITUTE
London, UK

WILEY

Registered Office
John Wiley & Sons, Inc., 111 River Street, Hoboken, NJ 07030, USA

Editorial Office
111 River Street, Hoboken, NJ 07030, USA

For details of our global editorial offices, customer services, and more information about Wiley products visit us at www.wiley.com.

Wiley also publishes its books in a variety of electronic formats and by print-on-demand. Some content that appears in standard print versions of this book may not be available in other formats.

Library of Congress Cataloging-in-Publication Data
Names: American Institute of Chemical Engineers. Center for Chemical Process Safety, author.
  Title: Bow ties in risk management : a concept book for process safety / CCPS, in association with the Energy Institute / Center for Chemical Process Safety of the American Institute of Chemical Engineers, and Energy Institute, London, UK. Other titles: Bow ties in risk management
Description: Hoboken, NJ : John Wiley & Sons, Inc. : American Institute of Chemical Engineers, 2018. | Series: Process safety guidelines and concept books | Includes bibliographical references and index. | Identifiers: LCCN 2018033748 (print) | LCCN 2018035050 (ebook) |
  ISBN 9781119490388 (Adobe PDF) | ISBN 9781119490340 (epub) | ISBN 9781119490395 (hardcover)
Subjects: LCSH: Chemical plants--Safety measures. | Risk management. | Organizational learning.
Classification: LCC TP150.S24 (ebook) | LCC TP150.S24 B69 2018 (print) | DDC 660/.2804–dc23
LC record available at https://lccn.loc.gov/2018033748

Cover Design: Wiley
Cover Images: Diagram courtesy of CCPS; Background © QtraxDzn/Shutterstock

Printed in the United States of America.

SKY10051356_071823

CONTENTS

LIST OF TABLES

LIST OF FIGURES

ACRONYMS AND ABBREVIATIONS

AIChE	American Institute of Chemical Engineers
ALARP	As Low As Reasonably Practicable
API	American Petroleum Institute
ATP	Authorized To Proceed
BOP	Blowout Preventer
CCPS	Center for Chemical Process Safety (of AIChE)
COMAH	Control of Major Accident Hazards (UK Regulation incorporating most of the EU Seveso Directive requirements)
CSB	Chemical Safety Board (US)
DNP	Do Not Proceed
ETA	Event Tree Analysis
ESD	Emergency Shutdown
EI	Energy Institute
EU	European Union
FMECA	Failure Modes, Effects and Criticality Analysis
FRAM	Functional Resonance Analysis Method
FTA	Fault Tree Analysis
HAZID	Hazard Identification Study
HAZOP	Hazard and Operability Study
HOF	Human and Organizational Factors
HSE	Health, Safety and Environment
HSE	Health and Safety Executive (UK)
IADC	International Association of Drilling Contractors
IOGP	International Association of Oil & Gas Producers
IPL	Independent Protection Layer
ISO	International Standards Organization
KPI	Key Performance Indicator
LOPA	Layer of Protection Analysis
LOTO	Lock Out Tag Out (part of Permit to Work)
LPG	Liquefied Petroleum Gas
MAE	Major Accident Event
MOC	Management of Change
MOPO	Manual of Permitted Operations
NFPA	National Fire Protection Association
NOPSEMA	National Offshore Petroleum Safety and Environmental Management Authority (Australia)
NORSOK	Norwegian Oil Industry Standards (Norsk Sokkels Konkuranseposisjon)
OSHA	Occupational Safety and Health Administration (US)
PHA	Process Hazard Analysis

P&ID	Piping and Instrumentation Diagram
PSA	Petroleum Safety Authority (Norway)
PTW	Permit To Work
QRA	Quantitative Risk Assessment
RBPS	Risk Based Process Safety
SCE	Safety Critical Element (also Safety or Environmental Critical Element or Equipment)
SIL	Safety Integrity Level (as per IEC 61508 / 61511 standards)
SIMOPS	Simultaneous Operations
SOOB	Summary of Operational Boundaries
STAMP	Systems Theoretic Accident Model & Processes

GLOSSARY

Terms in this Glossary, where relevant, match the online CCPS Glossary of Terms for Process Safety.

ALARP
As Low As Reasonably Practicable – a term used to describe a target level for reducing risk that would implement risk reducing measures unless the costs of the risk reduction in time, trouble or money are grossly disproportionate to the benefit. In bow tie analysis, it is a performance-based standard used for determining whether appropriate barriers have been put in place such that residual risk is reduced as far as reasonably practicable.

Barrier
A control measure or grouping of control elements that on its own can prevent a threat developing into a top event (prevention barrier) or can mitigate the consequences of a top event once it has occurred (mitigation barrier). A barrier must be effective, independent, and auditable. See also **Degradation Control.** (Other possible names: **Control, Independent Protection Layer, Risk Reduction Measure**).

Barrier Type
These are categories of a barrier. The purpose of defining a barrier type is to clarify its operational mode and to make transparent the case where only one type (e.g., active human) is relied on exclusively. Active barriers must contain the three elements of detect-decide-act.

•Passive Hardware
A barrier system that is continuously present and provides its function without any required action.

•Active Hardware
A barrier system that requires some action to occur to achieve its function. All aspects of the barrier detect-decide-act functions are achieved by hardware or software.

•Active Hardware and Human
The barrier detect-decide-act aspects are achieved by a mix of hardware, software and by at least one necessary human action.

•Active Human
The barrier detect-decide-act aspects are all achieved by humans. Some interaction with hardware will be necessary but the functions are predominantly human.

•Continuous Hardware	The barrier function is achieved by some continuous action.
Bow Tie Model	A risk diagram showing how various threats can lead to a loss of control of a hazard and allow this unsafe condition to develop into a number of undesired consequences. The diagram can show all the barriers and degradation controls deployed.
Consequence	The undesirable result of a loss event, usually measured in health and safety effects, environmental impacts, loss of property, and business interruption costs. Another possible name: **Outcome**. The magnitude of the consequence may be described using a Risk Matrix
Critical Barrier	An optional designation, sometimes required by companies or regulators, which identifies a subset of barriers that are designated to be more significant in risk control. The designation can assist prioritization of the barrier in terms of inspection, testing, maintenance and training. In principle, all barriers in a bow tie diagram are important and need an ongoing management process to ensure their effectiveness.
Dashboard	A simplified management diagram displaying KPIs or metrics (both leading or lagging) considered important in achieving the organization's safety, environmental or commercial objectives. Barrier status could be a key element to be displayed on a dashboard.
Degradation Factor	A situation, condition, defect, or error that compromises the function of a main pathway barrier, through either defeating it or reducing its effectiveness. If a barrier degrades then the risks from the pathway on which it lies increase or escalate, hence the alternative name of escalation factor. (Other possible names: **Barrier Decay Mechanism, Escalation Factor, Defeating Factor**).
Degradation Control	Measures which help prevent the degradation factor impairing the barrier. They lie on the pathway connecting the degradation threat to the main pathway barrier. Degradation controls may not meet the full requirements for barrier validity. (Other possible names: **Degradation Safeguard, Defeating Factor Control, Escalation Factor Control, Escalation Factor Barrier**).
Dike	Synonymous with bund. A passive barrier describing a secondary containment system around a tank, the walls of which act as the primary containment.

Hazard	An operation, activity or material with the potential to cause harm to people, property, the environment or business; or simply, a potential source of harm.
HAZOP	Hazard and Operability Study. A systematic qualitative technique to identify and evaluate process hazards and potential operating problems, using a series of guidewords to examine deviations from normal process conditions.
Human Factors	A term with both ergonomic and organizational implications. A discipline concerned with designing machines, operations, and work environments so that they match human capabilities, limitations, and needs. Human Factors is also the discipline used to describe the interaction of individuals with each other, with facilities and equipment, and with management systems. This interaction is influenced by both the working environment and the culture of people involved.
Impaired	Any degree of degradation of barrier performance from its intended function (i.e., partially available, not available, unknown status, etc.).
Incident	An event, or series of events, resulting in one or more undesirable consequences, such as harm to people, damage to the environment, or asset/business losses. Such events include fires, explosions, releases of toxic or otherwise harmful substances, and so forth.
Independence	The condition that no significant common mode of failure exists that would degrade two or more barriers simultaneously in an incident pathway.
LOPA	Layer of Protection Analysis. An approach that analyzes one incident scenario (cause-consequence pair) at a time, using predefined values for the initiating event frequency, independent protection layer failure probabilities, and consequence severity, in order to compare a scenario risk estimate to risk criteria for determining where additional risk reduction or more detailed analysis is needed.
Main Pathway Barrier	A barrier that lies along the direct route from a threat to the top event or from the top event to a consequence. (Another possible name: **primary barrier**).

MAE	Major Accident Event (MAE). A hazardous event that results in one or more fatalities or severe injuries; or extensive damage to structure, installation or plant; or large-scale, severe and / or persistent impact on the environment. In bow ties MAEs are outcomes of the top event. (Other possible names: **major accident, major incident**).
Metadata	Information about other information. In the barrier context, the base information would be the barrier name and description; metadata would be the collection of other data relating to the barrier.
Mitigation Barrier	A barrier located on the right-hand side of a bow tie diagram lying between the top event and a consequence. It might only reduce a consequence, not necessarily terminate the sequence before the consequence occurs (Other possible names: **Reactive Barrier, Recovery Measure**).
MOPO	Manual of Permitted Operations. An operational management diagram derived from bow ties that maps all required barriers that must be functional before a defined activity can be carried out. Impaired barriers must be repaired or replaced with an equivalent alternative before the activity can be carried out. (Other possible name: **Summary of Operational Boundaries – SOOB**).
Multi-Level Bow Tie	An advanced approach that extends the standard bow tie to show deeper level degradation controls that support degradation controls from themselves degrading. The first level of build-out beyond the standard bow tie is termed Extension Level 1. Additional extension levels are possible. (See **Standard Bow Tie**).
Pathway	A bow tie arm on which barriers or degradation controls are located. A Main Pathway is an arm connecting the various threats to the top event, or the top event to the various consequences and these contain barriers. (Alternative term: **Prevention Pathway** or **Mitigation Pathway**). Arms connecting degradation factors to a main pathway barrier are termed **Degradation Pathways** and these contain **Degradation Controls**.

Performance Standard	Measurable statement, expressed in qualitative or quantitative terms, of the performance required of a system, equipment item, person or procedure (that may be part or all of a barrier), and that is relied upon as a basis for managing a hazard. The term includes aspects of functionality, reliability, availability and survivability.
Prevention Barrier	A barrier located on the left-hand side of bow tie diagram and lies between a threat and the top event. It must have the capability on its own to completely terminate a threat sequence. (Other possible names: **Proactive Barrier**).
Process Hazard Analysis	An organized effort to identify and evaluate hazards associated with processes and operations to enable their control. This review normally involves the use of qualitative techniques to identify and assess the significance of hazards. Conclusions and appropriate recommendations are developed. Occasionally, quantitative methods are used to help prioritize risk reduction.
Process Safety Management	A comprehensive set of policies, procedures, and practices designed to ensure that barriers to episodic incidents are in place, in use, and effective.

The term is used generically in this document and is not restricted to the scope and rules of OSHA 29 CFR 1910.119 (frequently referred to as Process Safety Management or PSM). It is often aligned with the CCPS Risk Based Process Safety (RBPS) Guideline or the EI PSM Framework. |
| **RAGAGEP** | Recognized and Generally Accepted Good Engineering Practices (RAGAGEP) – a US regulatory requirement. They are the basis for engineering, operation, or maintenance activities and are themselves based on established codes, standards, published technical reports or recommended practices or similar documents. RAGAGEP details generally approved ways to perform specific engineering, inspection or asset integrity activities, such as fabricating a vessel, inspecting a storage tank, or servicing a relief valve. |
| **Risk Matrix** | A tabular approach for presenting risk tolerance criteria, typically involving graduated scales of incident likelihood on the Y-axis and incident consequences on the X-Axis. Each cell in the table (at intersecting values of incident likelihood and incident consequences) represents a particular level of risk. |

Risk Register	A regularly updated summary of potential major accident events over a facility life cycle, with an estimate of risk contribution and the barriers needed to achieve that level of risk. The risk register can be developed from facility PHA studies.
Risk Assessment	The process by which the results of a risk analysis (i.e., risk estimates) are used to make decisions, either through relative ranking of risk reduction strategies or through comparison with risk targets.
Safety I / II	A transition in safety thinking proposed by Hollnagel from where humans are regarded primarily as a source of errors in process safety (Safety I) to where humans are regarded as contributing more to ongoing safety successes (Safety II).
Safety Critical Element	Any part of an installation, plant or computer program whose failure will either cause or contribute to a major accident, or the purpose of which is to prevent or limit the effect of a major accident. Safety Critical Elements are typically part of barriers. In the context of this book, safety includes harm to people, property and the environment. (Other possible names: **Safety and Environmental Critical Element, Safety Critical Equipment**).
Safety Critical Task	A task where human or organizational factors could cause or contribute to a major accident, or where the purpose of the task is to prevent or limit the effect of a major accident, including: • initiating events; • prevention and detection; • control and mitigation, and • emergency response. Safety Critical Tasks are typically part of barriers.
Safety Integrity Level (SIL)	A relative level of risk reduction provided by a safety function, or to specify a target level of risk reduction. In simple terms, SIL is a measurement of performance required for a safety instrumented function (SIF). Defined in the IEC 61511 standard.

Standard Bow Tie   The basic bow tie showing hazard, top event, threats and consequences, with prevention and mitigation barriers, and optionally degradation pathways containing degradation controls supporting the main pathway barrier against identified degradation threats. (See also **Multi-Level Bow Ties**).

Swiss Cheese Model   A model of accident causation developed by James Reason. It represents a system of safety barriers depicted as slices of cheese with holes. In this model, the slices of cheese represent the safety barriers and the number and size of the holes an indication of the vulnerability of the barrier to fail.

Threat   A possible initiating event that can result in a loss of control or containment of a hazard (i.e., the top event). (Other possible names: **Cause, Initiating Event**).

Top Event   In bow tie risk analysis, a central event lying between a threat and a consequence corresponding to the moment when there is a loss of control or loss of containment of the hazard.

The term derives from Fault Tree Analysis where the unwanted event lies at the 'top' of a fault tree that is then traced downward to more basic failures, using logic gates to determine its causes and likelihood.

ACKNOWLEDGMENTS

The committee structure for this concept book differs from other CCPS books in that this was a joint project done in full collaboration with the Energy Institute. In addition, the contribution of the European Commission Joint Research Centre Major Accident Hazard Bureau is gratefully acknowledged. The American Institute of Chemical Engineers (AIChE) and the Center for Chemical Process Safety (CCPS) express their gratitude to all the members of the Bow Ties in Risk Management Subcommittee and their member companies for their generous efforts and technical contributions. Similarly, the EI acknowledges its Bow Ties in Risk Management Subcommittee, and to its Technical Partner and Technical Company Members for co-sponsoring the development of this concept book.

The authors from DNV GL and CGE Risk Management Solutions are also acknowledged, especially the principal authors Dr. Robin Pitblado and Paul Haydock, with additional inputs from Tatiana Norman, Jo Everitt, Amar Ahluwalia, Chris Boylan, and Ben Keetlaer.

Many of the figures in this concept book have been created in software, either from Thesis (ABS Group) or BowTieXP (CGE Risk). This contribution is acknowledged. Details on the software are provided in Appendix A.

PROJECT TEAM MEMBERS:

CCPS

Kiran Krishna	Shell	Project Team Chair
Timothy McGrath	ex Chevron	Project Team Vice-Chair
Americo Carvalho Neto	Braskem	
Umesh Dhake	CCPS Asia Manager	
Martin Johnson	BP	
Mark Manton	ABS Group	
Ron McLeod	Ron McLeod Ltd	
Darrin Miletello	Lyondell Basell	
Sudhir Phakey	Linde Gas	
Keith Serre	Nexen	
Ryan Supple	ConocoPhillips	

Thiruvaiyaru Venkateswaran	Reliance Industries	
Stephanie Wardle	Husky Energy	
Danny White	Ex-BHP Billiton	
Charles Cowley	CCPS Staff Consultant	Project Manager

Energy Institute

Mark Scanlon	Energy Institute	Project Team Co-Chair
Donald Smith	ENI	
Dennis Evers	Centrica	
Rob Miles	Hu-Tech	
Rob Saunders	Shell	

European Commission Joint Research Centre Major Accident Hazards Bureau

Maureen Wood
Zsuzsanna Gyenes

Before publication, all CCPS and EI books are subjected to a thorough peer review process. CCPS and EI gratefully acknowledge the thoughtful comments and suggestions of the peer reviewers. Their work enhanced the accuracy and clarity of this concept book.

Peer Reviewers:

San Burnett	BHP Billiton
Palani Chidambaram	Du Pont
Chris Devlin	Celanese
Scott Haney	Marathon Oil
Ed Janssen	Ed Janssen Risk Management Consulting
Bob Johnson	Unwin
Steve Lewis	Risktec
Don Lorenzo	ABS Group
Sian Miller	Newcrest Mining
Bradd McCaslin	Shell
Eric Wakley	Shell
Jack McCavit	JLM Consulting
Mary Metz	Director of Water Resource Policy Alberta
Louisa Nara	CCPS

Cathy Pincus	Exxon Mobil
Jan Pranger	Krypton Consulting
Karla Salomon	Chevron
Hans Schwarz	BASF
John Sherban	Systemic Risk Management Inc.
Mike Snyder	Dekra
Jeff Thomas	PII
Martin Timm	Praxair
Jan Windhorst	WEC Inc
Tracy Whipple	BP
Stuart King	EI HOFCOM and Tripod Foundation
Sam Daoudi	EI Process Safety Committee
Trish Kerin	IChemE Safety Centre
Sam Mannan	MKO Process Safety Center, Texas A&M University
Ian Travers	Ian Travers Ltd (ex Deputy Director Chemicals Regulation, HSE)
Mike Nicholas	Environment Agency
Mike Wardman	Health & Safety Laboratory (HSL)
Patrick Hudson	Independent Consultant, Emeritus Professor, Delft University

ONLINE MATERIALS ACCOMPANYING THIS BOOK

Although the bow tie figures in this book are shown in black and white and reduced in size to enhance readability, some of them are available in color and larger size in an online register.

To access this online material, go to:

www.aiche.org/ccps/publications/BTRM.aspx

Enter the password: BTRM2018

PREFACE

CCPS and EI Introduction

The American Institute of Chemical Engineers (AIChE) has been closely involved with process safety and loss control issues in the chemical and allied industries since the 1970s. AIChE publications and symposia have become information resources for those devoted to process safety and environmental protection.

AIChE created the Center for Chemical Process Safety (CCPS) in 1985 after the disasters in Mexico City, Mexico, and Bhopal, India. The CCPS is chartered to develop and disseminate technical information for use in the prevention of major chemical incidents. The Center is supported by around 200 chemical process industry sponsors that provide the necessary funding and professional guidance to its technical committees. The major product of CCPS activities has been a series of books to assist those implementing various elements of a process safety and risk management system. To complement the longer, more comprehensive *Guidelines* series and to focus on more specific topics, the CCPS extended its publication program in the last few years to include a 'Concept Series' of books. This book is part of the Concept Series.

The Energy Institute (EI) is the chartered professional body for the energy industry, developing and sharing knowledge, skills and good practice towards a safe, secure and sustainable energy system. The EI was set up in 2003 as the result of a merger between the Institute of Petroleum (IP) and the Institute of Energy (InstE). EI supports over 23,000 individuals working in or studying energy and 250 energy companies worldwide. The EI provides learning and networking opportunities to support professional development, as well as professional recognition and technical and scientific knowledge resources on energy in all its forms and applications.

The EI's purpose is to develop and disseminate knowledge, skills and good practice towards a safe, secure and sustainable energy system. It informs policy by providing a platform for debate and scientifically-sound information on energy issues. In fulfilling the EI's mission, its Technical Work Program addresses the depth and breadth of the energy sector, from fuels and fuels distribution to health and safety, sustainability and the environment. This program provides cost-effective, value-adding knowledge on key current and future issues affecting those operating in the energy industry, both in the UK and internationally. For further information, please visit http://www.energyinst.org.

Bow Ties in Risk Management Concept Book

CCPS has been at the forefront of documenting and sharing important risk assessment methodologies for more than 30 years. It has published well-known guidelines on hazard identification, chemical process quantitative risk assessments, Layer of Protection Analysis (LOPA), and facility siting. This concept book continues that tradition with a focus on a specific qualitative risk assessment methodology – bow tie barrier analysis.

Barrier-based risk assessment has been applied to process safety risks for over two decades and increasingly frequently through the use of bow tie diagrams. Bow tie barrier analysis focuses on assessing barriers for the prevention and mitigation of incident pathways, especially related to major accidents. Bow tie diagrams examine potential major accidents by diagrammatically mapping the hazards and threats that may lead to an event and the potential undesired consequences, including most importantly, all the barriers and degradation controls in place to reduce the risk. Bow tie diagrams can assist with barrier management, the analysis of risk reduction, and the assessment of barriers in place. They provide a powerful means to communicate complex process safety information to staff, contractors, regulators, senior management, the public, and other stakeholders.

The increasing use of bow ties to communicate risks and barriers has led the CCPS Technical Steering Committee to charter a project committee to develop this concept book for *Bow Ties in Risk Management.* The Energy Institute (EI) and European Commission Major Accident Hazards Bureau were collaborating partners with CCPS on this project. To gather input from many experienced sources, CCPS invited representatives from many chemical and petroleum companies, trade associations, and regulators involved in the field of process safety, as well as other key stakeholders or subject matter experts to participate in this committee's activities. The Energy Institute joined the project to share the knowledge of its members and particularly to provide additional focus on the human factors aspects of bow ties.

Well-constructed bow tie diagrams, which are clear and easy to communicate, may give the impression that they are easy to create. This is not the case. Too often bow ties are created with structural or other errors that detract from their value. The aim of this concept book is to equip the novice or even experienced reader with the requisite skills and knowledge in order to develop quality bow ties.

While there is currently a reasonable degree of consensus on how to handle technical matters in bow ties, the same is not true for Human and Organizational Factors (HOF). Chapter 4 addressing human factors in bow tie analysis is the product of a sub-committee representing a wide range of experience in the practice of human factors in the process industries, including both industrial and regulatory backgrounds. The sub-committee considered and critically evaluated how human factors issues are represented in current approaches to bow tie modeling. This group recognized the need for simplicity and clarity in bow ties as implemented, but also that oversimplification can lead to an incorrect understanding of how human factors actually contribute to safer operations. The approach described here addresses the critical role that people play in barrier systems, with the wide range of HOF that need to be managed effectively for barriers to be as robust as they reasonably can be – all with the aim of preventing barriers being degraded or defeated by 'human error'. Current approaches to bow tie modeling rarely capture the complexity of the human contribution to barrier systems and may not recognize the range of factors that need to be managed to mitigate the risk from 'human failure'. A multi-level bow tie method is proposed to capture these fully.

Therefore, even experienced bow tie practitioners may see changes to preferred terminology and will find novel material on HOF in this concept book. The committee believes that following these ideas will enhance the value, quality and consistency of bow ties produced, thus contributing to the goal of enhanced safety.

CCPS and EI encourage companies, regulators and other key managers of process risks around the globe to consider adopting and implementing the suggestions contained within this book.

1

INTRODUCTION

1.1 PURPOSE

The purpose of this concept book is to establish a set of practical advice on how to conduct bow tie analysis and develop useful bow tie diagrams for risk management. It describes the intended audience, gives directions on how to use the concept book and provides a basic introduction to the method, which is expanded in the following chapters. It explains the rationale for developing bow tie diagrams and how they fit into an overall risk management framework.

CCPS concept books address newer techniques in process safety that have not yet become accepted standard practice or where there is not yet industry consensus on approach. In EI publications, concept books are termed Research Reports in its series of technical publications. Concept books introduce these valuable tools in a simple and straightforward manner. CCPS and EI encourage the use of this concept book to aid the industry in developing better quality bow tie diagrams with a consistent methodology and preferred terminology for their use. Implementation of the methodology outlined in this CCPS / EI book should improve the quality of bow tie analysis and bow tie diagrams across an organization and industry.

1.2 SCOPE AND INTENDED AUDIENCE

This concept book provides practical advice on how to develop bow tie diagrams and in their use. This will help to:

- ensure consistent use of methodology and terminology;
- establish a valid approach to defining hazards, top events, threats, and consequences;
- establish criteria for barriers and degradation controls linked to degradation factors;
- identify common errors which may occur when constructing bow ties;
- provide a method to incorporate human and organizational factor issues in bow ties;
- provide guidance on how bow ties can be used for risk management purposes through the effective depiction of barriers;
- discuss basic and advanced uses of bow tie diagrams; and
- review an overall strategy for barrier management.

The intended audience for this book is primarily anyone involved with or responsible for managing process safety risks, although the concepts within the book are applicable to all bow tie risk management practices and not limited to process safety (e.g., for other safety and environmental applications and Enterprise Risk Management). It is designed for a wide audience, from beginners with little to no background in barrier management, to experienced professionals who may already be familiar with bow ties, their elements, the methodology, and their relation to risk management.

The origin of bow ties and their main use to date has been in managing process safety risks particularly relating to major accidents. They have been applied in the chemical / petrochemical and oil and gas industries as well as other industries (such as maritime, aviation, rail, mining, nuclear, and healthcare). However, the logic and approach described here may also be used to manage strategic risks, financial risks, risks of losing critical sales, etc. As these applications are less common, the examples in this book are focused on safety risks but also include human health, environmental impact, asset damage, and reputation loss.

Several software tools are available to aid in the development of the diagrams. The level of detail displayed using software tools can be complete or partial depending on the audience needs. This book does not endorse any particular software tool; however, Appendix A provides a summary of several widely available software tools known to the authors at the time of publishing.

1.3   ORGANIZATION OF THIS CONCEPT BOOK

This book is organized in a way that follows the logical flow of constructing bow ties and then conducting bow tie barrier analysis. Several examples are used to demonstrate this advice; however, these often only cover parts of a bow tie under discussion. More detailed, complete examples are found in Appendices B and C. The examples relate to the topics in each chapter and provide a story line concerning the development and use of bow ties. A summary of the content of each chapter is provided below.

Chapter 1 – Introduction
- Purpose and scope;
- Introduction to the bow tie concept;
- Linkage between bow ties, fault trees, and event trees.

Chapter 2 – The Bow Tie Model
- Define terminology and elements of the bow tie diagram;
- Illustrate robust and weak examples of bow tie elements;
- Define and discuss the types of barriers, including criteria for validity and quality.

Chapter 3 – Bow Tie Development

- Discuss the process of bow tie development including their initial development in team workshops;
- Discuss common errors and quality checks during development of bow tie diagrams.

Chapter 4 – Addressing Human Factors in Bow Tie Analysis

- Discuss how to include human and organizational factors in bow tie diagrams;
- Show how human factors can be addressed using a basic approach, but also introduce the concept of a multi-level bow tie which provides an extended analysis with greater utility, albeit with some complexity;
- Discuss metrics for human and organizational factors.

Chapter 5 – Basic Use of Bow Ties

- Discuss common uses of bow tie tools in analyzing barriers and identifying safety critical elements and tasks.

Chapter 6 – Management of Bow Ties

- Discuss use of bow ties as part of a barrier management strategy utilizing a lifecycle approach;
- Discuss links between bow ties and management system elements (e.g., Management of Change, maintenance, training, audits).

Chapter 7 – Additional Uses of Bow Ties

- Discuss the application of bow ties as a communication tool and to help demonstrate ALARP;
- Illustrate how bow ties are used to aid in decision making for various activities in an organization and for risk management;
- Show how real-time bow ties can interface with an organization's management system.

Appendices

- Appendix A – Software Tools; provides a table listing various software tools available to develop bow ties and their capabilities;
- Appendix B – Case Study for a pipeline; provides an example of a full bow tie with an emphasis on technical threats;
- Appendix C – Case Study for Multi-Level Bow Ties; provides an example bow tie incorporating human and organizational factors and demonstrating the concept of multi-level bow ties.

1.4   INTRODUCTION TO THE BOW TIE CONCEPT

The oil and gas industry has achieved a very impressive improvement in occupational safety. The fatality rate within the IOGP member companies has declined by an order of magnitude over the past 20 years, Figure 1-1 (IOGP, 2015).

However, the reduction in major process accidents has been less impressive than for occupational safety (Pitblado, 2011) and insured losses due to major accidents in the oil and gas and process industries have not reduced in the last 30 years (Marsh, 2016). Current risk approaches have tended to focus more on demonstrating design safety and less on maintaining operational safety.

The development and appropriate use of bow tie barrier diagrams have the potential to significantly improve process safety. They do this by focusing on the operational aspects, clearly highlighting all important safety barriers, helping in the assessment of barrier adequacy, communicating this visually to all staff and contractors, and providing a framework to continually monitor the effectiveness of these barriers. Bow ties can also be used in the design phase to test the adequacy and relevance of barriers and if additional barriers and degradation controls are required.

Once constructed, the bow tie purpose is best used to support risk management and risk communication. The bow tie diagram can provide a clear graphical representation of the output of the risk assessment and management process (threats, consequences, barriers and degradation controls) which is readily understood by people at all levels – from operational personnel and senior managers, to regulators, and to members of the public. The bow tie illustrates both the prevention barriers, which stop the top event from occurring, and the mitigation barriers, which reduce the consequence severity should the top event occur. For the full unmitigated consequence (i.e., major accident event), all of the barriers along the relevant pathway between the threat and the consequence must

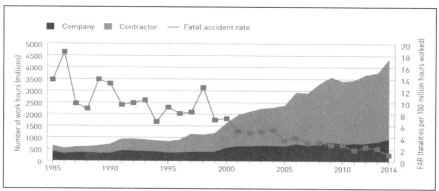

Figure 1-1. Fatal Accident Rate vs Total Hours Worked (Global Data)

fail or be degraded. Other factors not related to barriers may also contribute to the magnitude of the consequence (release orientation, wind direction, etc.).

Major accidents rarely result from a single failure, but rather from multiple barrier failures, which collectively result in a loss of control situation that further escalates to significant consequences. Common mode failures can be relevant in multiple barrier failures (e.g., underfunding or delayed maintenance can affect many barriers simultaneously). The bow tie approach helps the operational and maintenance teams to focus on barriers and the degradation controls which are relied upon to maintain their effectiveness. The effective management of all barriers is a key aspect of risk-based process safety management (CCPS, 2007).

1.4.1 Reason's Swiss Cheese, Models of Accident Causation and Bow Ties

James Reason (1990 and 1997) developed the idea of the 'Swiss Cheese Model' of system failure presented in Figure 1-2. The model builds on the principles of 'defense in depth,' with slices of Swiss cheese representing protective layers (i.e., barriers) preventing hazards from being realized and allowing consequences to happen. Reason observes that barriers are never 100% effective and each has unintended intermittent weaknesses. The holes in the cheese slices represent degradation factors (i.e., reductions in effectiveness or reliability) in individual parts of the system and are continually varying in size and position in all slices. For a major accident to occur, holes in the Swiss cheese need to align allowing for an 'accident trajectory' so that a threat passes through all of the holes in all of the defenses (i.e., barriers) leading to a failure or major accident. It also shows that if one barrier fails, then subsequent barriers are challenged. Using this model, a risk management strategy is successful when barriers are managed to ensure that they perform as intended at all times throughout the life cycle of the facility. When compared to the Swiss cheese model, bow ties add structure and give a better representation of the barriers associated specifically with multiple threat and consequence legs.

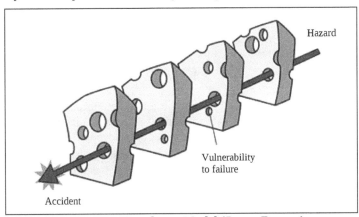

Figure 1-2. Swiss Cheese Model (James Reason)

Assumptions are frequently made, based on the visual structure of the representation, that bow ties assume a linear, event-driven model of technical systems and how they fail, i.e. a linear model of accident causation. It has been argued for some time that this type of model is inadequate as a means of understanding the dynamics of modern complex socio-technical systems or the ways they can lead to loss (Leveson, 2011; Hollnagel, 2012), particularly where there is both tight coupling and complex interaction between system elements (Perrow, 1999). However, adopting the concepts and structures of bow tie analysis need make no assumptions about the mechanisms and processes that lead to incidents (McLeod and Bowie, 2018). Reason's Swiss Cheese model appears linear in the time sequence in which barriers operate; development of the holes - i.e. the mechanisms of causation - can however proceed through either linear or non-linear processes (Hudson, 2014). Similarly, bow tie analysis, and the understanding of barriers, failure mechanisms and controls that it can generate, is neutral in terms of any underlying model of accident causation.

The bow ties and the model presented here of barriers and barrier degradation and control do not assume any particular mechanisms that might lie on the path between threats and the top events and consequences they can lead to. There is no reason why a bow tie model should not be based on non-linear analyses such as HAZOP (CCPS, 2008a), STAMP (Leveson, 2011) or FRAM (Hollnagel, 2012). In essence, rather than asking what could go wrong, and how to prevent it, the focus of bow tie analysis is on what needs to go right, and how to assure it (McLeod and Bowie 2018).

1.4.2 History and Regulatory Context of Bow Ties

It is generally accepted that the earliest mention of the bow tie methodology appeared in the ICI (Imperial Chemical Industries) course notes of a lecture on hazard analysis given at The University of Queensland, Australia in 1979, but how and when the method found its exact origin is not completely clear. Shell is acknowledged as the first major company to fully integrate the bow tie methodology into its business practices, and by the end of the 1990s, the approach became a common method within many other companies.

A large EU research project, called ARAMIS, investigated the use of bow tie risk assessment methodology in the framework of the Seveso II Directive. This project is described in a special issue of the Journal of Hazardous Materials (Salvi & Debray, 2006), which covers several aspects of bow ties, including their use and success for communication and as a means for organizational learning.

Regulatory regimes and standards are increasingly embracing the barrier concept as a means to assist in the management of risks during the operational phase and to communicate this effectively. In the process and oil and gas industries, the following list provides examples of the current major regulatory and industry bodies and how they are adopting a barrier-based approach:

- The American Petroleum Institute (API) – Recommended Practices 96, 65-2 and 90 are based on a barrier approach for offshore operations;

- UK Health and Safety Executive – guidance on safety management of major hazard industries (UK HSE, 2013a) recognizes bow ties as a tool or model for the barrier based approach. Following guidance from the UK, COMAH Competent Authority (SEPA, 2016a) has been extended to support assessment of environmental major accidents. As a consequence of this, many COMAH sites in the UK have included bow ties in their safety reports;

- The European Commission, under the 'Safety of Offshore Oil and Gas Operations Directive' – has put in place a set of rules to help prevent accidents as well as respond effectively, which requires a risk assessment and the identification of barriers;

- International Association of Drilling Contractors (IADC) – has developed a new WellSharp program, which includes a barrier-based approach to redefine well control training and assessment;

- The National Offshore Petroleum Safety and Environmental Management Authority (NOPSEMA), Australia – under Guidance Notes, N04300-GN0271, states that a barrier based layer of protection analysis may be required to prevent or mitigate hazards;

- The International Association of Oil & Gas Producers (IOGP), Reports 456 and 544 – recognizes the importance of barrier management and discusses methods for maintaining and keeping barriers up to date; and

- The Petroleum Safety Authority (PSA), Norway, 'Principles for Barrier Management in the Petroleum Industry' guidance document – focuses on barriers, barrier performance and barrier management as a means to reduce risk of accidents.   PSA conducts audits based on barrier management.

1.4.3 What Bow Ties Address

Bow ties are a useful tool to assist analysis of process and non-process industry risks. The general focus of bow ties in the process industry is towards Major Accident Events (MAEs – see Glossary for definition) as staff need to understand how these may occur and the barriers and degradation controls deployed to prevent them. While this implies large scale events with serious consequences, some companies choose to develop bow ties for less serious events, including occupational safety events such as 'fall from height'. This can be beneficial for risk management as it uses a common format; however, every bow tie requires some significant effort to create and ongoing efforts to communicate and remain current. Companies limiting bow tie applications to MAEs can reduce the total effort by using simpler methods for less serious events. Therefore, whether to limit bow ties to MAEs is a choice companies can make. Early applications were for large refinery, petrochemical or offshore facilities, but they also have important uses in small- and medium-size enterprises. They can also be applied to other industrial processes such as chemicals handling, transportation, storage facilities, and medical applications. Bow ties can support a structured approach to risk assessment for facilities that do not have P&IDs and so hamper conventional PHA (HAZOP) studies (e.g., mining, steel and other metal working industries). The emphasis is also different. HAZOP is used to identify hitherto unknown failure scenarios in complex systems, while bow ties demonstrate the barriers deployed for these scenarios. Bow ties also document how these barriers may fail and the processes and systems in place to prevent this from happening.

1.4.4 Key Elements of a Bow Tie

The bow tie diagram is shown in Figure 1-3 with the following elements:

1. **Hazard:** the bow tie starts with the hazard.
2. **Top Event:** the loss of control of the hazard.
3. **Threats** are depicted on the left side (customarily the prevention side) of the bow tie diagram.
4. **Consequences** of loss of control of the hazard are depicted on the right side (customarily the mitigation side) of the bow tie diagram.
5. **Prevention Barriers** on the left side of the diagram represent prevention barriers, which stop threats from resulting in the top event.
6. **Mitigation Barriers** shown to the right of the top event represent mitigation barriers, which mitigate the top event (i.e., reduce the scale of and possibly stop undesired consequences).
7. **Degradation Factors** can be applied to both prevention and mitigation barriers and these can lead to impairment or failure of the barrier to which they are attached.
8. **Degradation Controls** act to mitigate the Degradation factors, helping maintain the main pathway barrier at its intended function. Degradation controls can, but do not necessarily satisfy, the effective, independent, and auditable criteria for barriers.

Barriers and degradation controls are illustrated to show these as fundamental elements of the safety management system. Altogether, the diagram provides a holistic picture of the risk management system. The elements of the bow tie model are further discussed in Chapter 2.

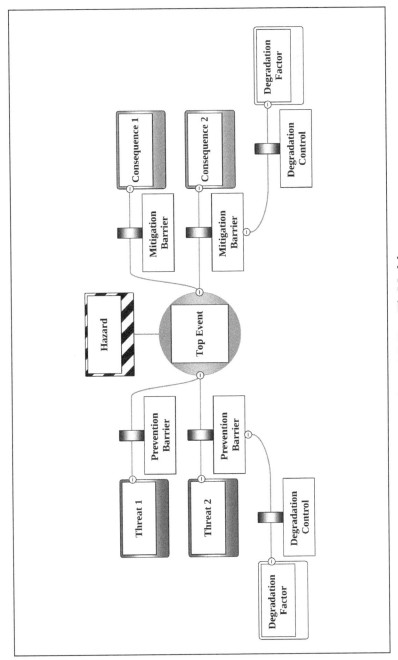

Figure 1-3. Bow Tie Model

1.4.5 Benefits of Bow Ties

Effective risk management requires a thorough understanding of all important incident pathways and implementation of applicable and reliable safety barriers that are in place to prevent and mitigate the risks. Well-drawn bow tie diagrams are visually obvious to a wide range of possible users, unlike numerical and complex results from other risk assessment methods such as QRA, LOPA, FTA and ETA which have different objectives and are more useful for specialists. For non-experts, the workforce, engineers and managers, bow ties provide a readily understandable and simple visualization of the relationships between the causes of events that lead to loss of control of the hazard, the potential consequences resulting from such events, the barriers preventing the event from occurring, and then the mitigation measures in place to limit its consequence. Most commonly, they are used when there is a need to demonstrate how hazards are controlled and to illustrate the direct link between risk controls and elements of the management system.

In order to develop a bow tie, it is necessary to first identify those hazards requiring such analysis. Most companies involved in hazardous activities have an existing process safety management system containing formal procedures and/or guidance for identification of potential hazards and assessment of risks. Bow tie analysis is not a methodology for identification of hazards; for this refer to 'Guidelines for Hazard Evaluation Procedures, 3rd Edition' for guidance (CCPS, 2008a). However, once hazards have been identified, the bow tie method can be applied to graphically demonstrate the management of the hazard and provide a framework for demonstrating effective control.

Bow tie analysis overlaps with LOPA in that they both examine barriers on pathways for major accidents. This is discussed more fully in Chapter 7.

To summarize, the main reasons to create and implement bow tie diagrams are to:

- provide a systematic analysis of the barriers along threat and consequence pathways that can prevent or mitigate a Major Accident Event. This can support the design process, or the operational or maintenance activities to raise awareness and understanding of the barriers in place and the role of individuals in operating or maintaining barriers;

- provide a cumulative picture of risk through the visualization of the number and types of barriers and degradation controls and their condition. This can support the identification and prioritization of actions to strengthen degraded barriers and degradation controls; and

- provide a structured process where identified hazards, threats and consequences can be related in cause and effect scenarios, and to assist in the development and understanding of how unwanted events can occur.

More advanced uses of bow ties for operational safety, integrating human factors, barrier management, incident investigation, and supporting dashboards are provided in Chapters 4 to 7.

Bow ties can demonstrate the link between controls and the management system, specifically those relevant to the management of risks (e.g., safety critical elements, critical roles / tasks / activities). Once all the barriers and degradation controls are identified, then an ongoing management program is required to ensure that the barriers are maintained at their specified effectiveness and that degradation controls are as robust as they reasonably can be. The bow tie diagram helps to ensure that this happens by clearly indicating the barriers in place, and the degradation controls associated with maintaining those barriers. The identification of safety critical elements and systems can then support the establishment of performance standards. These concepts are developed in Chapter 2.

Bow ties help with accountability by documenting barrier owners who can track the condition of their barriers and how often any are involved in incidents. Bow ties can identify critical tasks and hence link these to competency requirements for procedures and administrative controls, and to required training and development for employees. They can also be used to enhance incident investigations to identify patterns of barrier failures, and to assist organizational learning.

There are different depths of treatment that are possible with bow ties. Some companies focus on the communications aspects only, while others attempt to develop real-time monitoring of barrier conditions using bow ties as the basis. Organizations need to consciously determine their strategy for incorporating bow ties into their process safety management system and to keep this up to date.

There can be drawbacks with the bow tie approach and this is acknowledged. If bow ties are poorly drawn showing too many non-applicable barriers, this can give users a false sense of security. This can be because degradation factors are shown as barriers or there may be dependencies between barriers that suffer from a common cause fault. Such barrier diagrams also make the communication task more difficult. It is also possible that PHA scenarios may not be accurately transposed into the bow tie diagram and important scenarios are omitted. Suggestions later in this book assist those creating bow tie diagrams from making these types of errors.

Bow ties have multiple functions in the realm of risk management, much more than simply communication of risk, although this is a key use. They address major accident risks during the operational phase in a format that is useful for operational and maintenance personnel. Bow ties can also be used in the design phase, although use in the operational phase dominates as many important operational barriers are not defined until that stage. Basic uses of bow ties are discussed more fully in Chapter 5, while additional uses appear in Chapter 7.

1.4.6 Linkage between Bow Ties, Fault Trees, and Event Trees

The bow tie methodology is sometimes described as a combination of two existing risk analysis tools, Fault Tree Analysis and Event Tree Analysis. These techniques are presented with examples in Guidelines for Chemical Process Quantitative Risk Assessment (CCPS, 2000). Optically rotating an FTA clockwise (as these are usually drawn vertically) and connecting it to an ETA (usually drawn horizontally) creates a bow tie connected at the top event, but there are significant differences. FTA and ETA can be quantitative, while bow ties are qualitative and focus on creating a visually simple representation of hazards, threats, top events, barriers, degradation factors and controls, and consequences.

A FTA describes an incident (top event) in terms of the combinations of underlying failures that can cause them and connecting these with AND or OR gates, while bow tie analysis places no visible Boolean logic between barriers. However, a bow tie 'barrier' would be represented in a Fault Tree as a 'demand on barrier' AND 'barrier fails', so the bow tie can include Boolean logic but it is usually hidden as it is unnecessary for the user. Different threats in bow ties are in effect OR gates directing into the top event (i.e., the top event may be caused by threat 1 OR threat 2 OR threat 3, etc.). In FTA the top event can be either the loss of control or loss of containment as is the case in a bow tie, or it can go further towards the undesired consequence. Some of the rule set for FTA also applies to bow ties. This includes the requirements for independence.

ETA requires identification of safety functions or barriers. It also considers hazard promoting factors excluded from bow ties, such as wind direction, early or late ignition, etc. ETAs have branches with 'barrier fails' and 'barrier succeeds' as the two distinct outcomes. In an ETA, 'barrier success' terminates the event; that arm is discarded or the branch determines a different outcome, for example, immediate ignition 'flash fire' or delayed ignition 'explosion'. The bow tie does not display arms that terminate, only arms that continue on to a consequence.

Since FTA and ETA may be quantitative and since the bow tie represents these visually, it may be possible in the future that researchers may be able to extend bow tie analysis to quantitative applications, but that is not addressed in this concept book. Some bow tie software tools do permit quantitative LOPA analysis using a bow tie as a starting point.

1.5   CONCLUSIONS

The bow tie methodology has been in use for over 20 years, and its focus has been towards the management of major accidents in multiple industries. It builds on the Swiss cheese model of accident causation and illustrates diagrammatically how threats can act on hazards to generate a loss of control, which may result in undesired consequences. In the bow tie diagram, prevention barriers are located on the left side and mitigation barriers are located on the right side. A well-drawn bow tie clearly shows all barriers that can prevent the top event from occurring or mitigate the consequences (see Appendix B for an example bow tie). Additional information can be incorporated in the bow tie to show degradation factors acting on barriers and the degradation controls that are implemented to maintain barriers at their intended functionality. Other information such as barrier effectiveness, type, ownership, and status can also be displayed.

A bow tie diagram is a powerful tool for communicating how the control of major accident hazards is achieved. It is focused on the operational phase although it has design applications as well, and since it is diagrammatic and non-quantitative, it can be used more easily for risk communication than quantitative risk tools.

Ongoing process safety management requires that all barriers perform at their expected level of effectiveness throughout a facility lifetime and that the degradation controls that are relied on to support barriers are actually in place. A bow tie diagram can be used to help monitor and provide prioritization to maintain the integrity of all barriers and degradation controls. The diagrams then become an integral part of the facility process safety management program.

2

THE BOW TIE MODEL

2.1 BOW TIE MODEL ELEMENTS

This chapter describes the characteristics of each element of the bow tie. Advice is provided for the correct formulation of the elements, with both good and poor examples offered. The bow tie model contains eight elements, which are described in this chapter; formal definitions appear in the Glossary. Listed in the typical sequence of building the bow tie, these elements are: (1) hazard, (2) top event, (3) consequences, (4) threats, (5) prevention barriers, (6) mitigation barriers, (7) degradation factors, and (8) degradation controls. This is shown in Figure 2-1. To differentiate this bow tie from a later described advanced use, multi-level bow tie, this form is called the standard bow tie.

Within the bow tie methodology, the hazard is an operation, activity or material within an organization that has the potential to cause harm. The top event is the moment in time when control over the hazard is lost. The consequences on the right are unwanted outcomes to which the top event can lead. The threats on the left side are the possible causes that can trigger the top event. The barriers on the left side, called prevention barriers, have the capability to prevent the threat from occurring, or to stop the threat from escalating to the top event. The barriers on the right, the mitigation barriers, either stop the top event from leading to the consequences or mitigate the consequences. For clarity, all the barriers attached directly to threats or consequences can be called 'main pathway barriers'. Barriers may not function as intended and may become degraded. Degradation factors are those factors that cause impairment, and may be attached to any barrier. The final element in the bow tie are degradation controls, which are intended to stop or reduce the likelihood that the degradation factor impairs the main pathway barrier. Degradation controls contribute to maintaining barrier effectiveness by preventing degradation of the barrier, but often may not themselves meet the required criteria for full barrier validity.

Some other industries prefer the terms 'proactive' rather than prevention, and 'reactive / recovery' rather than mitigation. In the context of the process industry, the terms prevention and mitigation work well for MAEs, but for example in aviation, recovery may be preferable on the right-hand-side. This is because the consequence of a crash is not mitigated, but it may be possible to recover from the loss of control.

The following sections discuss each element in more detail. Examples are given for each element, while full bow ties are provided in Appendices B and C.

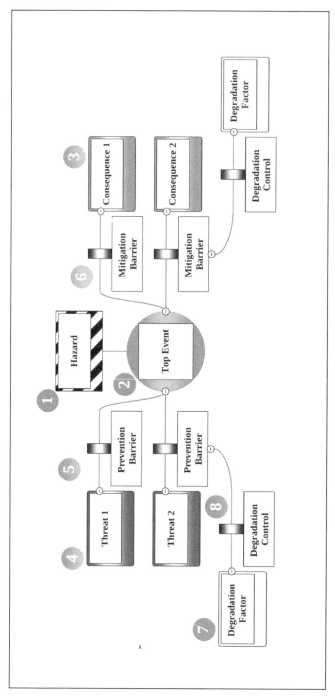

Figure 2-1. Standard Bow Tie Showing all the Basic Elements

2.2   HAZARD

2.2.1 Hazard: Characteristics

The 'hazard' is an operation, activity or material with the potential to cause harm. It is shown on the diagram to provide clarity to the reader as to the source of risk. Hazards are part of normal business and are often necessary to run an operation.

Some examples of hazards are toxic materials, high pressure gases, rotating equipment, flammable liquids in atmospheric storage tanks, loading a tanker truck, manufacturing polyethylene, or processing hydrocarbons that are flammable and under pressure. A hazard example is shown in Figure 2-2.

The hazard has the important function of defining the scope for the whole bow tie. Generic hazards can lead to generic bow ties and thus the hazard should be specific. This tends to add value because it increases the level of detail in the rest of the bow tie. Examples of well-defined and poorly defined hazards provided in Table 2-1 and Table 2-2 will make this clearer.

Two types of details that can be relevant to include in the hazard:

A) *Situational context.*   Situational information can be essential to comprehend the type of hazard, such as geographical location, business unit or phase of operations, and other concurrent activities.

B) *An indication of scale.*   The scale that is involved can also provide important information. What is the capacity of the tank? Under how much pressure is the chemical substance?

2.2.2 Formulating the Hazard

Hazards would normally be identified in a PHA process (e.g., HAZID or HAZOP). Additionally, the Hazard Checklist in ISO 17776 (2016) can provide guidance on a wide range of potential hazards. Although it is an offshore standard, it provides good general guidance for onshore facilities as well.

| Gasoline stored in tank | 'Gasoline' alone would be too generic a description, but by linking the gasoline to the storage tank it is possible to identify risk scenarios. |

Figure 2-2. Gasoline Storage Tank Hazard

Describe the hazard in its controlled state. Hazards should be formulated in a controlled state. A hazard description should be 'transporting fuel in a truck from A to B' and not 'fuel truck explosion'. A hazard describes a potentially harmful substance / process / activity and not the loss of control of the hazard (this is the 'top event') or the actual harm that can lead from that process (these are the consequences). A useful check in defining the bow tie hazard correctly is to ask, "Is the hazard as described part of our normal business?"

Add detail to the hazard to determine scope and desired bow tie detail level. If the hazard is well defined this will support more useful bow ties. The guideline 'Be specific' will apply to almost every element, but it is probably most important for the hazard, as the level of detail set in the hazard will influence the level of detail in the rest of the bow tie. The hazard box on the bow tie diagram cannot show all the details of the hazard but the specifics should be documented. The correct level depends on the scope and purpose of the bow tie. Operators might for example have a single transportation bow tie that would cover their operations whereas a road transport company may have many driving-based bow ties for their different modes of transportation. In the case of the process industries, specify the material being processed or handled. The material properties and processing / storage conditions will provide insight into the potential consequences.

2.2.3 Hazard Examples

Some well-worded and poorly-worded example hazards are provided in Table 2-1 and Table 2-2. In the following examples, both process industry and normal life instances are provided, as some readers of bow ties may better understand the non-process examples. Generally, one- or two-word hazard descriptions may be too short and not convey adequate detail about the potential hazard.

Table 2-1. Well-Worded Hazard Examples

Hazard	Comment – why this is well worded
Pressurized propane storage in sphere	Normal operational state is defined and volume in sphere will be known to those using the bow tie
Driving a tanker on the highway	Driving a tanker on the highway is a normal requirement to get from A to B. This in itself is not a problem, but it does have the potential for loss of control.
Drilling in a formation with hydrocarbons under pressure	Drilling in a rock formation with hydrocarbons is part of normal business for oil and gas companies, but does have the potential to cause harm (e.g., blowouts).

Table 2-1. Well-Worded Hazard Examples, continued

Hazard	Comment – why this is well worded
Processing hydrocarbons containing H2S gas	Hydrocarbons have the usual flammable properties; the H2S gas is an additional toxic hazard that points to wider safety issues. Since these hazards are different with some possible differences in barriers, this might justify two bow ties with one focusing on flammable hazards and one on toxic hazards.
Working at height (>2m) on formwork	Working at height is a well-known hazard, specifying the height provides additional detail.
Transporting people to and from a work site via helicopter	The activity of flying in a helicopter to a work location is well defined.
10 tons of pressurized chlorine in a storage tank	The hazard of chlorine is further specified by providing the maximum tank contents.

Table 2-2. Poorly-Worded Hazard Examples

Hazard	Comment – why this is poorly worded
Chlorine	This is too vague – is it product chlorine in small cylinders, in piping, or in the main storage tank?
Uncontrolled fire	'Uncontrolled fire' is a consequence; it is not a part of normal business. However, 'fighting a fire at a chemical facility' is a possible hazard, as it is an accepted part of business for the firefighting unit.
H2S	The hazard does not properly set the scope, nor identify the scenario that will be analyzed. The bow tie will be entirely different depending on whether we are drilling into a formation containing H2S, smelting iron with H2S as by-product or working in sewers where H2S is present.
Ignition	This is part of an incident sequence; it is associated with the top event 'loss of containment' of the hazard 'hydrocarbons in the process'.
Control system failure	This can be a threat, a top event, or a barrier failure, depending on the context. It does not specify the actual hazard – perhaps high pressure process fluid.
Derailment	'Derailment' is not a good description of a hazard, because it is not a part of normal business; it is in fact a top event. A better hazard would be transport of crude oil by train.

2.3   TOP EVENT

2.3.1   Top Event: Characteristics

Given a well-selected hazard description, the next step is to define the top event in the center of the diagram.

The top event is the moment when control over the hazard or its containment is lost, releasing its harmful potential. While the top event may have occurred, there may still be time for barriers to act to stop or mitigate the consequences. An example top event linked to the prior hazard example is shown in Figure 2-3.

It is possible to identify multiple top events for one hazard – control can be lost over the hazard in different ways. Therefore, one hazard can result in multiple bow tie diagrams. For example, the hazard 'working at height' can result in two top events 'dropped object' and 'person falls from height'. This will lead to two bow tie diagrams with different top events, but the same hazard.

The term top event derives from the fault tree, which has similarities with the left side of the bow tie model. This would also then be the starting point for the event tree analysis that leads to all the possible consequences. Fault and event tree linkages were discussed in Section 1.4.6.

2.3.2   Formulating the Top Event

Describe how / what control is lost. The top event describes an event in which control of the hazard is lost. Common generic top events are loss of containment,

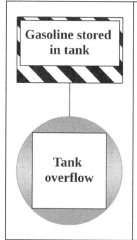

	The hazard (from the prior section) now links directly to the top event – which is defined as a tank overflow event.
Gasoline stored in tank	
Tank overflow	The tank overflow is the top event. It is not a consequence as there is no serious loss yet apparent. However, possible consequences (safety or environmental outcomes) may follow and there are several mitigations that can lessen their effect (e.g., quick shutoff, dikes, ignition controls).

Figure 2-3. Example of a Hazard and Top Event

loss of separation, loss of stability (e.g., of a floating drill rig) or loss of control (e.g., of a chemical reaction).   In process safety applications dealing with hydrocarbons, the most common top event is loss of containment.

Give an indication of scale if possible. As with the hazard specification, it is often good practice to quantify the top event. Thus, rather than just 'hydrocarbon leak', it might be better to differentiate rupture and small leak as many of the barriers and consequences will be different. Adding scale details to the top event can be useful for the discussion of threats, barriers and consequences as everyone will have the same scenario in mind, and this assists creating a well-constructed bow tie diagram. Two separate top events would mean two separate bow ties; a more efficient alternative would be to define small and large consequences associated with different leak sizes using a single bow tie.

The top event should not be a consequence. A common error in defining a top event is to choose a consequence with damage or harm rather than a loss of control event. To prevent this error, check the top event and ask, "Is this the loss of control or is this a consequence?"   As an example, an employee fatality is generally the end of the chain of events with multiple mitigation barriers preceding it. This is not a good top event, as there will be only a left side of the bow tie and no right side.

Choosing the best top event. Selecting a suitable top event may be easy, especially for loss of containment events, or loss of control that leads obviously to an MAE. However, many analysts have found experience and consultation are invaluable in choosing less obvious top events. The cost of selecting a second-best top event - perhaps not incorrect but right or left of the focus of your study - can be a skewed bow tie that fails to address and treat the real risks the team is attempting to control.

A useful check can be to identify at least two threats and two consequences for the bow tie (the next steps in building the bow ties). If these threats and consequences cover the areas of interest to the organization and to the department that will own the bow ties, the right moment in time / in the process has been chosen as the top event for this department's study.

In formulating the hazard and top event, the analyst should always be thinking "Is this top event too narrow so that we will need several diagrams to cover the risks surrounding this asset or operation? Can we do the same analysis using one bow tie rather than several? Or is it too broad, and should we split it up to several bow ties?" A test is to ask: "How many threats and consequences can we build for this top event?" If it is only one or two, the top event may be too narrow. If it is more than ten, perhaps it is too broad (or by its nature it is possible that there are many valid threats). The balance between detail and economy is influenced by the intended audience, the purpose and criticality of the study, any history of incidents, etc. However, there can be cases where a single threat or consequence is worthy of analysis if their magnitude is sufficiently large.

As an example, the technically correct top event of 'tank overflow' in Table 2-3 does not cover all loss of containment scenarios, and for those, other bow ties might be required. To avoid having other diagrams, the analyst might consider replacing 'tank overflow' with 'loss of fuel from the tank' so that other possible threats (e.g., corrosion leading to leaks through the tank floor, structural failure, impact, valve or fitting failure, etc., as well as overfilling) can be included on the one bow tie. However, there have been so many tank overflow events, that it is probably best to have a bow tie specific for overflow. A 'too broad' example might be a top event of 'process fluid loss of containment'. Different barriers and hence bow ties might be appropriate for vapor or liquid releases or if the material were flammable or toxic.

2.3.3   Top Event Examples

Table 2-3 and Table 2-4 provide several examples of combinations of well and poorly-worded hazard and top event examples.

Table 2-3. Well-Worded Top Event Examples

Hazard	Top Event	Comment – why this is well worded
Gasoline stored in a tank	Tank overflow and gasoline spill onto dike floor	The hazard links directly to the loss of containment event. Multiple consequences are possible which will be explored on the right side of the bow tie.
Driving a tanker on the highway	Loss of control of the tanker	In this case, loss of control is literal – losing control of the tanker is the top event.
Drilling into formation containing hydrocarbons under pressure	Loss of well control Influx of hydrocarbons	The loss of well control can be due to either an influx of hydrocarbons into the well or a loss of drilling fluids into a permeable formation. Since the barriers are different, two bow ties are appropriate: one for the influx of hydrocarbons and the other for loss of drilling fluids. The top event makes clear which is the causal mechanism.

Table 2-3. Well-Worded Top Event Examples, continued

Hazard	Top Event	Comment – why this is well worded
Loads suspended by a crane	Dropped object	The dropped object is the loss of control over the lift. It leads to several possible undesired consequences, but with multiple mitigations; hence this is a good top event. It may be appended by 'or swinging loads' or be changed to 'loss of control of the load'.
Hydrocarbons under pressure in processing unit X	Loss of containment to atmosphere	The top event defines a loss of containment event inside the processing unit X. This is sufficiently clear to enable review of barriers for ignition controls, gas detection, fire response, etc., and excludes loss of (primary) containment to a flare system.

Table 2-4. Poorly-Worded Top Event Examples

Hazard	Top Event	Comment – why this is poorly worded
Gasoline stored in a tank	Tank overflow and major dike fire	This top event combines the actual top event with one of the possible consequences. It bypasses all the various mitigation barriers that aim to prevent ignition and reduce the consequence of a major fire.
Gasoline stored in a tank	Corrosion of the tank	'Corrosion of the tank' can be a good top event, but is not correct for this hazard. 'Corrosion of the tank' does not describe how control is lost over 'storing hydrocarbons in an atmospheric tank'. 'Corrosion of the tank' describes one of the threats that can lead to loss of control over the hazard (e.g., loss of containment).
Driving a tanker on the highway	Crashing into a tree	'Crashing into a tree' is not a good top event. A crash is not a way we lose control over our hazard, but the unwanted result of losing control over our hazard. We can identify our real top event by asking 'What was the initial loss of control that led to the crash?'

2.4   CONSEQUENCES

After the top event is defined, the next step is to determine the consequences. Given the shape of the bow ties with threats on the left and consequences on the right, the 'natural' approach might be to define threats first. However, changing the order and defining consequences before threats will often help the team later to define only the threats that acting on the hazard can lead to significant consequences. Some analysts do define threats first, so the sequence consequences or threats first is optional.

Normally the bow tie might consider safety, environmental, asset damage, and reputation losses, although this will depend on the mandate given to the team. Often there may be multiple consequences of one type. An example of this is due to ignition. Immediate ignition of flammable releases might lead to jet fires or pool fires, whereas delayed ignition may give rise to vapor cloud explosions. Different barriers are needed to address these separate outcomes.

An example hazard and top event leading to a consequence is shown in Figure 2-4.

2.4.1 Consequences: Characteristics

Consequences are unwanted outcomes that could result from the top event and lead to damage or harm. Generally, these would be major accident events, but lesser consequences can be selected if the aim is to map the full range of important safety and environmental barriers. One top event may have multiple consequences – but normally only important consequences would be developed to show the mitigation barriers, not trivial ones. Risk matrix information can also be displayed on the consequences to show the severity of loss or damage that a consequence might cause and its likelihood. This information is often available from the HAZOP or PHA study.

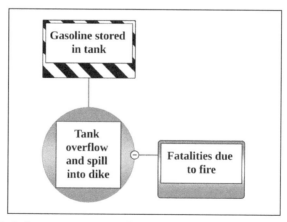

Figure 2-4. Consequence caused by the Top Event of Tank Overflow

2.4.2 Formulating Consequences

Consequences should be described as '[Damage] due to [Event]'. It is important to include the event leading to the damage, as different barriers can be required to stop or mitigate damage depending on the event leading to the damage.

'Fatalities due to fire' might, for example, call for different mitigation barriers than 'fatalities due to toxic gases', and 'environmental damage due to smoke' can require different barriers than 'environmental damage due to liquid spill'. Care should be taken to avoid being too specific in defining consequences, such as splitting injuries from fatalities. When reviewing a bow tie, if all the barriers are the same on different pathways, they could normally be combined, unless differences in the risk assessments are worth noting. It is generally easier to combine consequence events later than to split out consequences that have been found to be too generic. Some analysts use a visual shorthand to show only the differences in barriers when multiple consequence pathways are almost identical – this makes communication of the differences easier. However, this feature requires manual drawing as this option may not be present in software.

Some consequences may also become a threat on another bow tie (for example, 'dropped object impacts plant' may be an initiator to a process loss of containment) and this topic is addressed in bow tie chaining in Section 7.8.

In the UK and elsewhere, when formulating consequences, it is important to consider the expectations of local regulators and whether any local guidance on qualifying or quantifying damage has been produced. For example, within Europe the Seveso Directive highlights scale of consequences to differing receptors including people, the environment, and property. Further guidance has been published regarding environmental receptors (SEPA, 2016b) and experience has shown that understanding the expectations of regulators, through discussions with them, is important in framing any risk assessment work.

Care should be taken to avoid developing a consequence that does not flow directly from the top event. This may be a temptation to address an orphan consequence that otherwise might be missed. For example, in a tank overflow top event a consequence of 'internal tank explosion' would not be appropriate. It is better to develop a bow tie specifically for this consequence, with a suitable hazard, top event, and threats.

2.4.3 Consequence Examples

As with the other elements of a bow tie, consequences can be chosen which are good or poor, but generally selecting consequences is less prone to error than some other elements. Table 2-5 and Table 2-6 give one consequence for each top event although multiple consequences are likely and will be shown on the bow tie. For example, 'dropped object' could also have a safety consequence as well as asset damage.

Table 2-5. Well-Worded Consequences Examples

Top Event	One Consequence †	Comment – why this is well-worded
Tank roof sinks	Asset damage from full surface tank fire	The consequence links directly to the top event and will allow all the various mitigation barriers to be properly included. It is specific in the type of consequence.
Loss of control over the vehicle	Driver injury / fatality due to crash into object	This range of outcomes is also a suitable consequence. Since the mitigation barriers would be the same, it is sensible to combine both injury and fatality into one consequence.
Loss of well control	Major harm to marine wildlife due to oil pollution.	This consequence is acceptable as it defines the scale of environmental damage at least qualitatively. Most company risk matrices include categories ranging from minor through to catastrophic, so indicating scale is useful.
Dropped object	Impact damage and total loss of object that is dropped	This consequence is clear and directly results from the top event.

†: The bow tie will have multiple consequences leading from the top event.

Table 2-6. Poorly-Worded Consequences Examples

Top Event	One Consequence†	Comment – why this is poorly worded
Gasoline tank overflow	Environmental damage or Pollution	The consequence links directly to the top event but it is vague, and not specific as to the nature or severity of the environmental damage. Is the damage to land or water (small stream or large river?) or to specific species? Consequences should name the receptor affected. Inclusion of the scale is useful to design an adequate response from the mitigation barriers.
Loss of control over the vehicle	Crash barrier damage	This is a possible consequence, but it is likely to be unimportant compared to other consequences and might be better grouped (e.g., 'asset damage to car and road infrastructure').

Table 2-6. Poorly-Worded Consequences Examples, continued

Top event	One Consequence †	Comment – why this is poorly worded
Dropped object	Delay	This consequence is also too vague. If this is a heavy lift of a critical piece of infrastructure, then delay is an important consequence and some magnitude will be important, e.g., 'project delay for over 3 months'.
Loss of containment	Evacuation of the facility	A plant will be evacuated when a loss of containment of hydrocarbons escalates to a stage that recovery is no longer possible. The evacuation is however not the actual consequence but a barrier to prevent worse consequences such as multiple injuries.

†: The bow tie will have multiple consequences leading from the top event.

2.5   THREATS

2.5.1 Threats: Characteristics

Threats are potential reasons for loss of control of the hazard leading to the top event. For each top event there are normally multiple threats placed on the left side of the diagram (as in Figure 2-1), each representing a single scenario that could directly and independently lead to it. A more specific threat for the gasoline tank example in Figure 2-4 is shown in Figure 2-5.

When a team is brainstorming threats, a PHA or HAZID is often a valuable input as these document causes leading to major accident events, but not necessarily all potential causes. This is because the breadth of those studies is usually broader and the time per issue is necessarily less than in a bow tie session which tends to be focused on only a few top events. It is important to remember that the threat, if the pathway is not prevented, must lead to the top event. In addition, three categories are helpful to initiate discussion in identifying threats:

1) primary equipment not performing within normal operating limits (e.g., mechanical fault – pump seal failure),

2) environmental influence (e.g., overpressure due to solar heating of blocked in pipeline),

3) operational issues (e.g., insufficient personnel present to support all required human barriers during unit start-up).

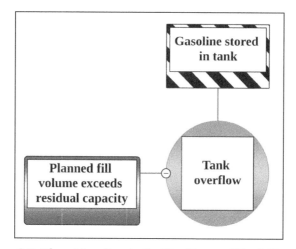

Figure 2-5. Threat Leading to the Top Event of Tank Overflow

Loss of containment is generally the result of one of the following issues:

- overfilling / underfilling;
- overpressure / under pressure;
- corrosion;
- stress / fatigue;
- incorrect flange torqueing;
- embrittlement;
- erosion;
- wear and tear;
- physical damage / impact; or
- subsidence / settlement / earthquake.

Besides using these categories as inspiration, it can also be helpful to ask the question as to why a certain procedure or protocol exists - there is usually a good reason. Probably it is meant to control something (a possible threat in a high-risk scenario, for example). This threat can then be added to the bow tie.

The use of 'human error' as a threat leading directly to a top event is generally not recommended as this commonly leads to structural errors in the bow tie as the barriers suggested are more often degradation controls. A structural error in a bow tie means that some important rule for bow tie construction has been violated. This may be any significant error in the eight elements of a bow tie (hazard, top event, threat, etc.), barriers not meeting the full validity requirements, or incorrect placement of degradation factor controls onto a main pathway representing them as full barriers. Structural errors can cascade as the first error can lead to further errors (an incorrect hazard can lead to an incorrect top event and this to incorrect threats, barriers and consequences). Experience shows human error is better treated as a degradation factor leading to impairment of a main pathway barrier. This is discussed in more detail in Chapter 4.

A frequent mistake is to exclude threats that will rarely lead to the top event because it is argued that there are already many prevention barriers in place to control this threat. The reason that these measures were implemented in the first place is to make these possible threats very low risk. Every credible threat should be added to facilitate decisions as to whether there are enough prevention barriers of sufficient strength in place to control the particular threat. Visualizing and recording the credible threats enables a more complete overview.

Where threats use identical barriers, these can be combined on a single threat pathway. The team may brainstorm 'mechanical failure – faulty flow meter', and if this requires the same barriers as 'mechanical failure – faulty level gauge', they can be combined. One threat leg is sufficient and it might be renamed 'mechanical failure – faulty level gauge or flow meter'.

2.5.2 Formulating Threats

Threats should have a direct causation and should be specific. A threat is direct when the causal relationship between the threat and the top event is clear without additional explanation. This helps to avoid misunderstandings during the making and communicating of the bow tie diagram. For example, direct threats for the top event 'loss of control of vehicle'' can be 'driving on slippery road' and 'reduced visibility'. 'Bad weather conditions' does itself not describe what will cause someone to lose control over their vehicle. The threat must be direct in causation but not in the sense of immediate time as it can occur years before the top event does (e.g., a latent threat such as a software programming fault may require a specific combination of inputs to reveal itself).

Identifying direct threats will often result in the inclusion of more specific barriers compared to indirect threats. For example, different barriers may be identified to control driving in bad weather conditions in general than to control driving on slippery road conditions or poor visibility specifically. Usually, generic threats lead to generic barriers, whereas specific threats lead to specific barriers. Specific barriers give more practical information leading to a better understanding of what should be done to prevent a threat acting on the hazard to produce a top event.

Although specificity is generally better, there is a balance to be struck. The best advice is to be pragmatic and choose a direct threat without being lost in semantic discussions on causality. The aim of this concept book is to demonstrate how to create a bow tie that is specific instead of abstract, which increases the quality and value of the bow tie.

Threats should be sufficient. Each threat itself should be sufficient to lead to the top event. If a threat can only cause the top event in combination with another threat, it is not sufficient in itself and therefore incorrect. When two threats need to occur together to lead to the top event, then these are called 'necessary' threats and should be reformulated into one independent threat. For example, 'incorrect close-out of isolation after maintenance' (note that this is actually a barrier failure, as discussed in the next section) and 'starting up operation' will not independently lead to the top event 'loss of containment'. Only when both occur at the same time can this lead to the top event. These threats could be reformulated to the one threat 'start up operation while valve is open'.

Even though one sufficient threat should be able to lead to the top event, in practice, the top event is not necessarily the result of only one threat. Many threats acting simultaneously could contribute to a top event – threats associated with corrosion and deviations in pressure can each result in vessel rupture as a top event, but if both mechanisms are present then the top event can occur with greater likelihood than either on their own.

Threats are not barrier failures. Formulating a threat as the failure of a barrier is one of the most frequent mistakes when constructing a bow tie diagram. A barrier failure on its own does not lead to a top event, because the barrier failure is a control that stops the actual threat from reaching the top event. Absence of a barrier means that the system relies on other barriers to protect against the threat. A failed barrier, such as a broken lock in a Lock-out Tag-out, has no 'energy' and does not initiate or accelerate an unwanted chain of events; by contrast, a threat, such as mechanical impact, actually contains the energy to lead to a top event. Not all threats have to contain energy – sometimes it is an unwanted condition or event such as a seal failure or incompatible software.

When barrier failures have erroneously been written as threats, these can usually be spotted by certain words such as 'lack of (...)', 'failure of (...)', 'absence of (...)', etc. Although these words can signal a barrier failure, they can also signal failure of primary equipment. Primary equipment are pieces of equipment that are part of the process, and failures of this equipment are valid threats (such as 'pump seal failure'). These are valid threats because each has the power, on their own, to lead to the top event. In contrast, equipment that has a purely safety related function is not a threat as failure of safety equipment is actually a barrier failure that is called on to act by some valid threat. However, in unusual circumstances, failures of safety equipment can become a threat in another bow tie (for example, the failure of a CO_2 fire suppression system is not a threat in a fire focused bow tie, but faulty actuation can result in asphyxiation and should be addressed in a different bow tie).

2.5.3 Threat Examples

Tables 2-7 and 2-8 provide several examples of well and poorly-worded threats.

Table 2-7. Well-Worded Threat Examples

Threat	Top Event	Comment – why this is well worded
Excess filling of tank	Tank overflow	The threat links directly to the top event without the need for other combination threats and it is a credible cause.
Excess speed for road conditions	Loss of control over the vehicle	The threat links directly to the top event. The hazard would be driving a vehicle.
Cementing failure in well	Loss of well control	A poor cement job can allow formation hydrocarbons to enter the well once the dense fluids above are removed.
Lifting unbalanced load	Dropped object	An unbalanced load on lifting equipment can cause a load to fall and thus is a direct cause of the top event.

Table 2-8. Poorly-Worded Threat Examples

Threat	Top Event	Comment – why this is poorly worded
Level gauge out of preventive maintenance cycle	Tank overflow	The threat is not a direct cause of tank overflow just because it is late on a preventive maintenance cycle. The threat is excess flow into the tank and the barrier is associated with operator vigilance using the level gauge.
Failure of anti-lock braking system (ABS)	Loss of control over the car	This is a safety system which has failed. It does not cause the top event on its own. A better threat would be a sudden burst tire.
High pressure well	Loss of well control	This could be a threat, but it is poorly worded. All wells increase in pressure at greater depths, so this is a normal condition. A better threat for this issue would be 'unexpected pressure increase in well'.
Wind during lift operation	Dropped object	Wind can lead to swinging load and ultimately to a dropped object, but the threat is too generic. A better threat would be 'Strong wind (>40mph)' as this is a much clearer indication of the challenge to the integrity of the lift.

2.6   BARRIERS

This concept book differentiates between barriers and degradation controls. Barriers appear on the main pathways (threat to top event or top event to consequence). Barriers must have the capability on their own to prevent or mitigate a bow tie sequence and meet all the validity requirements for a barrier to be effective, independent, and auditable. Degradation controls only appear on degradation pathways and serve as measures to support main pathway barriers against the degradation threat. These are discussed in Section 2.7. Degradation controls do not directly prevent or mitigate the event sequence, as that is the role of the main pathway barriers.

2.6.1   Barriers: Concept and Location on Bow Tie

Barriers can be physical or non-physical measures to prevent or mitigate unwanted events. Active barriers can differ with respect to the 'detect', 'decide', and 'act' components they contain and whether these components are performed by humans or executed by technology. The detection mechanisms of the barrier detect the current state and based on this state, a decision is made, and an action is performed if needed. The active barrier topic is addressed in further detail in Section 2.6.2.

Prevention and mitigation barriers will be treated in further detail in this section. Degradation controls against degradation factors are discussed in depth in Section 2.7. Figure 2-6 shows the location of prevention and mitigation barriers on a main pathway of a bow tie.

A barrier is placed on the bow tie diagram where it delivers its function or effect. In a bow tie with 'loss of containment' as top event and 'fire' as consequence, the barrier 'automatic firefighting system' is only effective after the top event and therefore appears on the right side of the bow tie diagram. A good question to ask is whether a barrier will prevent the top event, in which case it should be placed on the left – or whether it acts after the top event, in which case it appears on the right.

Barrier properties. These are similar to the characteristics defined in LOPA (CCPS, 2001) for an Independent Protection Layer (IPL); barriers must be effective, independent, and auditable (see more detail in Section 2.6.3). The barrier is the complete system fulfilling these criteria. Active barriers must contain all the elements of detect-decide-act (so 'gas detection' is not a barrier, but 'gas detection, logic solver, and ESD would be). It is a common error to list the elements of detect-decide-act as separate barriers rather than just one.

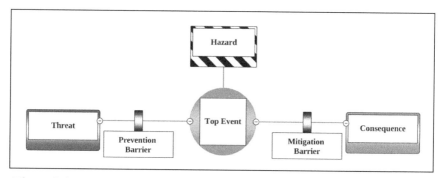

Figure 2-6. Bow Tie Showing Prevention and Mitigation Barriers on Either Side of the Top Event

Prevention barriers. A prevention or threat barrier is a barrier that prevents the top event from occurring. A key test for a prevention barrier is that it must be capable of completely stopping the top event on its own. This does not mean that it is 100% reliable, only that in principle it can prevent or terminate a threat sequence (for example, a properly sized pressure relief valve can prevent a top event of 'pressure vessel burst', but it can fail if the degradation control 'routine recalibration of relief valve' does not occur). It is noted that some in the process industry accept prevention barriers that only reduce the likelihood of the top event but not necessarily have the capability of preventing it. This view was reviewed by the CCPS / EI Committee and rejected due to the focus of this book on MAE events and their serious potential consequences. The reduced likelihood approach may be valid for less serious scenarios.

There are two main ways in which a prevention barrier can have effect: either to prevent the threat from occurring in the first place, or to stop an occurring threat from leading to the top event. The differentiation in function of the barrier can be useful, mainly to facilitate the brainstorming session regarding barriers. One can ask "Can we think of ways to prevent the threat?" followed by the question "Can we think of ways to stop the top event from occurring once the threat has taken place?"

Mitigation barriers. Mitigation barriers (on the consequence side of the bow tie) are employed after the top event has occurred and should help an organization prevent or reduce losses and regain control once it has been lost.

There are two main ways in which a mitigation barrier can have effect: either to stop the consequence from occurring (ignition prevention), or to reduce the magnitude of the consequence (detection, response, and ESD). A mitigation barrier can have a lower performance than a prevention barrier in that it may only mitigate, not terminate, a consequence. (A water curtain barrier may reduce toxic ammonia concentration but not eliminate it; similarly, an ignition control barrier only reduces the likelihood of ignition but does not eliminate this potential.)

Mitigation barriers also need to be effective, independent, and auditable; and if active, contain all the elements of detect-decide-act.

2.6.2 Barriers: Type and Characterization

Barrier type identifies the main operating characteristic of the barrier. While several classifications might be possible, it is suggested to use five types as suggested below. These first four are listed in the sequence of effectiveness, giving a hierarchy of control. The position of Continuous Hardware is dependent upon the specific system. More details with examples are given in Table 2-9.

- passive hardware;
- active hardware;
- active hardware + human;
- active human;
- continuous hardware.

There are other barrier type names in use; for example, some companies use Procedural (for Active Human) or Socio-technical (for Active Hardware + Human). However, the five names suggested here are more descriptive of the whole range of possible barrier types and easier for new staff to understand. In addition, labels such as Procedural might lead teams to nominate specific procedures as barriers, and since a procedure is just a piece of paper, this does not meet the requirements for a full barrier.

A common error is to categorize Active Hardware + Human barriers as Active Hardware. While the most visible elements of such a barrier consist of hardware (e.g. a gas detection and ESD system), this misses the human contribution to the decide aspect of detect-decide-act (e.g. to push the ESD button). The correct type is Active Hardware + Human. It only becomes Active Hardware if all three aspects of detect-decide-act are hardware, for example as in an automatic logic based shutdown system.

Active barriers must have separate elements of 'detect-decide-act', i.e., 'detect' a change in condition or what is going wrong, 'decide' what action is required to rectify the change and 'act' to stop the threat from progressing further (Figure 2-7). These three terms are sometimes called 'sensor', 'logic solver', and 'actuator' by some bow tie users, relating the terms to common elements of a safety instrumented system.

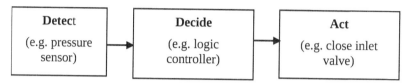

Figure 2-7. Detect-Decide-Act Model

If any of the detect-decide-act elements is missing from an active barrier, the barrier will not be able to stop the threat. For example:

1) A firefighting system could be perfectly designed for realistic fire scenarios, but it will not function if no 'detect' element is present to allow a person or controller to decide that the system is required and then to activate it.

2) A very good alarm (detect) is ineffective if it does not lead to a suitable response action. Thus, the barrier would be 'alarm and operator response'. Operator response includes both the decide and act elements.

3) An 'emergency shutdown valve' (act) on its own is not a barrier. The system must include 'detect and decide' elements or the barrier will not function.

If any element of detect-decide-act is missing, then showing this as a barrier gives a false sense of security by portraying a barrier that is not fully functional. Table 2-9 provides details on barrier types and how 'detect-decide-act' applies for active barriers.

Table 2-9. Barrier Types and Linkage to 'Detect-Decide-Act' Elements

Short Name	Barrier Type †	Description	Detect	Decide	Act	Examples
Passive	Passive Hardware	The barrier works by virtue of its presence.	N/A	N/A	N/A	Dike, blast wall, crash barrier, anti-corrosion paint
Active	Active Hardware	All elements of the barrier are executed by technology.	Technology (e.g., pressure sensor)	Technology (e.g., logic controller)	Technology (e.g., emergency shutdown valve)	Process control systems and Safety Instrumented Systems
Human ††	Active Hardware + Human (predominately hardware)	The barrier is a combination of human behavior and technological execution.	Technology (e.g., high-high level indicator and alarm)	Human (e.g., operator hears and responds to alarm)	Technology (e.g., emergency shutdown valve) OR Human (e.g., operator manually shuts valve)	Operator-activated ESD valve Gas alarm and decision by human to evacuate
	Active Human (predominantly human)	The barrier consists of human actions, often interacting with technology.	Human observation (e.g., operator walk around detects leak)	Human evaluation (e.g., decides to shut-down and isolate the equipment)	Human – but acting on technology (e.g., operator presses stop button or manually shuts a valve)	Operator detection and response (e.g., during structured walk arounds)
Continuous	Continuous Hardware	The barrier is always operating.	N/A	N/A	Technological	Ventilation system, impressed current cathodic protection

†Not all barriers can fit exactly within the barrier type model, particularly for mitigation barriers (e.g., ignition control is a blend of passive (electrical switch cubicles) and active hardware (shutdown of powered systems)).

††Active Human could be combined into Active Hardware + Human as even where all three aspects of detect-decide-act involve humans, some hardware is usually involved (e.g., in "act": a human presses a stop button to shut a valve); however, the distinction is helpful to teams to identify barriers that are predominantly human.

2.6.3 Barrier Properties

Each valid barrier should be effective, independent, and auditable. It must have the capability to completely stop the threat from leading to the top event or, if a mitigation barrier, significantly reducing or eliminating the consequence. Each barrier must be independent of other barriers on the pathway and the threat.

The CCPS book on LOPA (CCPS, 2015) extended the core attribute list for IPLs to seven attributes:

- independence;
- functionality;
- integrity;
- reliability;
- auditability;
- access security; and
- management of change

Companies can extend the three attributes suggested for barriers to the full list of seven for IPLs if they wish, but the three attributes already restrict many potential barriers. CCPS (2015) describes control measures that do not meet the criteria for a barrier.

Effective. A prevention barrier is described as 'effective' if it performs the intended function when demanded and to the standard intended, and it is capable on its own of preventing a threat from developing into the top event. A mitigation barrier is described as 'effective' if it is capable of either completely mitigating the consequences of a top event, or significantly reducing the severity.

Examples of common mistakes when representing effective barriers on a bow tie include:

- Referencing 'training' and 'competency' as barriers: these are degradation controls and would appear on a degradation pathway supporting the barrier to which they apply.
- Identifying incomplete barriers: e.g., 'fire & gas detection'. While these are important barrier elements, they do not constitute a complete barrier as they rely on other elements to completely stop the scenario from developing further. For this example, a complete barrier could be 'fire and gas detection, automatic logic controller (or human response to alarm) and ESD'.

Independent. Barriers should be independent of the threat and of other barriers on that pathway. For example, if the threat was loss of power and a barrier requires power to operate, then that would not be a permissible barrier in that pathway. A common mode failure occurs when one event causes two or more barriers to fail. Ideally, there should be no common mode failure possible for all the barriers in a pathway and they should satisfy the 'independence' property. Unfortunately, this is practically impossible. All barriers tend to have some commonality, either being maintained or operated by the same team, or even just being part of the same organization. In most situations, common mode failures that affect all barriers are not very likely. But realistic commonalities do exist, such as barriers that are reliant on the same critical service (e.g., electricity), maintenance based on the same maintenance management system, or the same person being responsible for or performing the procedural steps of multiple barriers.

There is a growing awareness that environmental aspects can affect barrier independence, for example, flooding. There are several examples where flooding has impacted multiple barriers and allowed top events to occur as this aspect was not considered sufficiently in the design. Facilities may start to include a barrier of preventive shutdown in the same way that Gulf of Mexico offshore oil facilities are shut down and evacuated when hurricane impacts are possible.

Although it is important to have as little common mode between barriers as possible, it is not necessary to remove barriers with some minor aspect of a common mode. As highlighted above it is often difficult to find barriers that have no common mode. Additionally, the barriers may have a common mode in one scenario (for example, in power outage), but work independently in other scenarios. Nonetheless, this risk of a plausible common mode failure should be managed by the addition of other barriers that do not have this common mode. Adding different types of barriers (such as active and passive, e.g., firewater system and firewalls) is advisable and usually can help avoid some general common mode failures.

Including multiple barriers which suffer from a common mode of failure on the same prevention or mitigation pathway creates an illusion of safety, so only include one of them. This is advisable if the common mode of two barriers is a likely reason of failure of the barriers. In the case of two barriers that rely on the same operator to push a button, the operator is a common mode and absence of the operator could be a likely reason for failure of both. When one barrier then fails because of the absence of the operator, the second barrier provides minimal additional security and should be removed from the bow tie. It would be useful to add a note to the diagram where barriers have been removed due to common mode failure issues. The use of degradation factors may highlight the susceptibility of different barriers on a single pathway to common mode failure.

Where serious common mode issues are identified then more detailed offline analysis may be required using reliability techniques, such as fault tree analysis or simulation.

Auditable. Barriers should be capable of being audited to check that they work. Formally, it could be that performance standards are assigned to the functionality of a barrier. For example, a performance standard for an ESD valve would ideally include 'periodic end-to-end testing', i.e., a signal is placed upon the detection device, the logic controller responds, and activates the end device, e.g., the ESD valve.

Degradation controls can also be auditable: in the example from Figure 2-11, an audit program can be put in place to verify that HAZOPs have been carried out and that they are of acceptable quality.

Number of barriers on each threat or consequence pathway. Grouping together equipment and tasks so that only 'effective, independent, and auditable' barriers are represented typically results in the number of barriers on the bow tie reducing to between one to five barriers on each threat or consequence pathway. Consequence pathways often have more barriers than prevention pathways. Keeping the number of barriers low has a major benefit in that the bow tie is more easily understood, and management and operations personnel do not gain a false sense of security believing that many barriers are in place when, in reality, several may not be independent, may be partial barriers, or may actually be degradation controls.

Bow ties do not explicitly represent the frequency with which threats are represented. This can mean there is no easy way to establish the risk associated with various pathways and hence perhaps the number of barriers required or criticality of barriers. This is more a task for LOPA or other risk assessment techniques. Bow tie teams assess risk by qualitative means. A comparison of decision making using LOPA and bow ties is given in Section 7.2.

It is partly for these reasons that quantitative targets for the number of barriers to employ are discouraged. If a team sees that a pathway might be short of a barrier according to some quantitative criteria, it might attempt to split a single barrier into two, put a degradation control onto the main pathway, or accept two barriers with a known common failure mode. Avoiding targets is better, and instead use a test of practicality, RAGAGEP or ALARP, to decide whether additional barriers are needed.

Splitting or combining threat and consequence pathways. It is important that a barrier is effective against all instances of the threat or consequence. This is described as a barrier having 'full coverage'. For example, the threat 'slippery road conditions' could have barriers that are relevant for slippery roads due to snow, but no barriers for slippery roads due to rain, giving users the false impression that all scenarios are treated. To include more specific barriers in the bow tie, such as using snow chains, it would be better to use a threat pathway for slippery roads due to snow and a second threat pathway for slippery roads due to rain. In general, consider splitting the threat (or consequence) into multiple, more specific threats (or consequences) when some barriers are only applicable in a special case of the threat (or consequence).

During barrier identification, it can also become clear that it is better to combine some of the threats or consequences. For example, two different threats may be found to have the same barriers. In such a case, showing the threats separately does not add new information, so it is better to combine them into a single threat. Splitting and combining are both part of the iterative process. Combining threats provides broader coverage and reduces the overall size of the bow tie diagram, supporting more effective communication.

Some organizations have combined consequences of different types (e.g., people and asset damage), on the grounds that most of the barriers are the same. However, this is not usually advisable as, for example, asset-damage mitigation barriers having no effect on people can give a false sense of risk control. In addition, management and reporting of the risks and strategies can be more complex. Some users address this by only showing differences between repeated similar consequence pathways, though this requires manual drafting as most current bow tie software does not support this functionality.

Splitting or combining pathways correctly requires judgement and is an iterative process. It is difficult to make the right choice for threats (and consequences) immediately, so it may be necessary to change things in light of deeper understanding of the barriers. Ultimately, there is no single 'right' answer. The resulting bow tie will be a function of the team developing it, their experience, priorities, and operating philosophies.

Barriers should be placed in time sequence of their effect. The most logical way of placing barriers in a diagram is in time sequence of their effect. The advice is thus to place the barriers in the order in which they are called upon, so it is clear when each barrier is needed. Often this means that design controls appear first (e.g., steel containment envelope, which includes the design criteria of material selection, pressure specification), followed by operational controls, then automatic trips, etc. An example is provided for one of the most commonly used threat lines for a loss of containment bow tie 'Operating outside of operating envelope' shown in Figure 2-8.

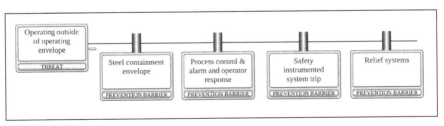

Figure 2-8. Demonstration of Time-ordered Barrier Sequence

Generally, the same barrier should not appear on both sides of the top event. However, this may occur for some special types of barriers which have both prevention and mitigation function. This is not common but some safety devices, such as a blowout preventer system on offshore wells, do have multiple functionalities: the annular rams have a prevention function and the shear rams have a mitigation function. This can make it difficult for teams to decide which side of the bow tie to place the BOP barrier.

While it is possible to sequence barriers differently, such as by criticality, effectiveness, or priority, this is not advisable. The problem is that it is not intuitive and therefore everyone who reads the bow tie needs to be made aware of what criteria have been used to sort the barriers. Adding metadata to the barriers can communicate this information in an efficient manner (see Section 2.6.4).

Barrier elements and grouping. To define barriers that adequately conform to the effectiveness and independence requirements, they frequently need to consist of a number of elements. A more detailed description could include all elements – both technical and human. An example might be a 'steel containment' barrier including all of the piping, vessels and pumps in a process plant. These items are all supported by a maintenance and inspection system though only particular areas of concern needing attention may be drawn out on the bow tie. These might be shown as safety critical tasks and frequently appear in degradation pathways supporting a main pathway barrier (see Section 2.7).

The Norwegian safety regulator (the PSA) issued a public guideline on barrier management (PSA, 2013, 2017) that identified three levels of barrier definition:

- barrier function – the task or role of the barrier (e.g., stop leaks, prevent ignition, reduce fire loads, ensure effective evacuation)
- the barrier system – the totality of the technical, operational and organizational elements whose collective function reduces the possibility of a specific error, hazard or accident to occur, or which limits its harm or disadvantages (e.g., asset integrity system, ESD shutdown system)
- barrier elements – technical, operational or organizational measures or solutions each of which plays a specific part in realizing the overall barrier function (e.g., gas detection, operator judgment, ESD valve).

The PSA document also shows performance standards as possible barrier sub-functions. In this concept book, performance standards are suggested as barrier metadata (Section 2.6.4).

The PSA nomenclature, which is gaining use in Norway, is shown in Figure 2-9. In this concept book, "barriers" usually correspond to PSA "barrier systems". For example, a high-pressure shutdown system will have barrier elements including a pressure sensor, logic controller and shutdown valve. Individual elements of a main pathway barrier can be spelled out in the barrier name or may appear as degradation controls against the degradation factor "high pressure

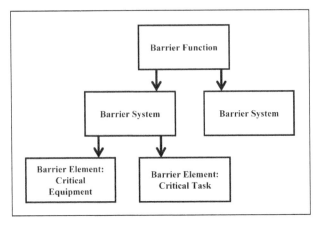

Figure 2-9. Barrier Hierarchy

shutdown fails to operate". Bow ties developed in accordance with the guidance in this concept book do not show the barrier function in the bow tie as it is usually clear from the threat or the top event. (In the example, the function might be 'stop excess pressure'.)

2.6.4 Metadata

When the bow tie contains all required basic elements (called a standard bow tie as shown in Figure 2-1), the result is an overview of the risks and how they are managed. This overview can become even more useful by adding additional information, or 'metadata', of which several different types are available depending on the nature of the various bow tie elements. Such metadata can usually be displayed or hidden by bow tie software, depending on the communication objective. The most common types of barrier metadata include: effectiveness or strength; condition; accountability and barrier type.   The preference of which metadata to display can differ during design and operation. For example, displaying effectiveness may be more important during design risk reviews, whereas condition is more important during operational risk reviews.

A full listing of typical metadata is provided in Section 5.3.3.

2.6.5 Barrier Examples

Barrier titles are important to communicate clearly the specific function of the barrier. Well-worded, short titles help to communicate the barriers deployed and for quality checking. The most common mistakes regarding barriers are:

1) displaying multiple barriers that are actually elements of a single barrier;

2) having barrier titles that are not informative;

3) placing barriers on the wrong side of the bow tie top event; and

4) including measures which are not barriers at all (e.g., degradation controls which belong on degradation factor pathways, e.g., training, competence).

Table 2-10 and Table 2-11 show examples of ways in which these errors tend to manifest themselves.

Table 2-10. Well-Worded Barrier Examples

Top Event – Threat / Consequence	Barrier	Comment – Barrier Type Descriptions
Tank overflow – Hydrocarbons affect environment	Mitigation: Dike	This is a passive hardware barrier as the dike is continuously present. It is somewhat of a simplification as dikes must have some way to drain rainwater, and if a drain valve is used this may be left open. This should be shown as a degradation factor for the dike.
Loss of control over the car – Driver impacts dashboard	Mitigation: Air bags	This is an active hardware barrier as the air bag system must detect when deceleration is above a critical threshold and then actuate an ignition device.
Loss of containment to water - Major environmental pollution event	Mitigation: Detect leak and deploy spill response equipment	This is a Hardware + Human barrier as it combines mechanical booms and boats with operator actions.

Table 2-11. Poorly-Worded Barrier Examples

Top Event – Threat / Consequence	Barrier	Comment – why this is poorly worded/placed
Tank overflow – Faulty level gauge	Prevention: Deploy foam protection	The barrier description is poor as it does not convey Detect-Decide-Act clearly. In addition, categorization as a prevention barrier is incorrect as deploying foam protection only occurs after the spill event, so this is a mitigation barrier. Foam does have a fire prevention function, but it is still a mitigation barrier as it acts after control of the hazard has been lost (i.e., after the top event).

Table 2-11. Poorly-Worded Barrier Examples, continued

Top Event – Threat / Consequence	Barrier	Comment – why this is poorly worded / placed
Loss of control over the car – Driving too quickly	Prevention: Crash barrier	The passive barrier description is good, but the placement is incorrect. It is a mitigation barrier because the crash barrier provides its function after control of the vehicle has been lost.
Loss of well control - Poor cementing job	Prevention: Blowout Preventer (BOP)	This is a difficult barrier as BOPs have multiple safety devices with both a prevention function and a mitigation function; the exact definition of loss of well control is also a factor (e.g., influx of hydrocarbons or uncontrolled blowout). Thus, the barrier should be renamed to make clear which part of the BOP is providing the prevention function (e.g., annular rams).
Dropped object – unbalanced load	Prevention: Watchman	This barrier, if it is just someone watching the activity, is not effective for prevention. It could be that the team meant that this person would verify that the crane safety systems were properly deployed and that a job safety assessment had been conducted. This would be an example of a useful control that is so poorly described that it conveys no useful information to the reader.
Loss of containment – Fire	Mitigation: Fire detection system	This barrier only contains a 'detect' component and no 'decide' or 'act' capability.
Leakage – Fire	Mitigation: Adhering to emergency response plan	Although this barrier is technically correct, it is a very generic barrier that could be placed on almost any bow tie. Consider whether this barrier is useful in the bow tie, given its goal and audience. A greater focus on fire response would be appropriate.

Table 2-11. Poorly-Worded Barrier Examples, continued

Top Event – Threat / Consequence	Barrier	Comment – why this is poorly worded / placed
Loss of containment – Seal failure	Prevention: Maintenance plan	The maintenance plan is not a measure that can stop the threat. A better barrier would be 'appropriate seal fitted to specification', and the maintenance plan is then a degradation control on the degradation factor line to ensure that the seal integrity is maintained.

2.7   DEGRADATION FACTORS AND DEGRADATION CONTROLS

2.7.1 Degradation Factors: Characteristics

Degradation factors and degradation controls are drawn in the bow tie diagram below the barriers to which they apply (see Figure 2-10). Controls along the degradation pathway are called degradation controls. The degradation factor is a condition that can reduce the effectiveness of the barrier to which it is attached. A degradation factor does not directly cause a top event or consequence, but since it degrades the main pathway barrier, the likelihood of reaching undesired consequences will be higher.

Degradation controls are described in Section 2.7.2. These controls frequently do not fully meet the criteria of a barrier (effective, independent, and auditable). Using the term "degradation controls" clearly indicates when they do not meet the barrier standard although note that some practitioners use the terms "degradation barriers" or "safeguards".

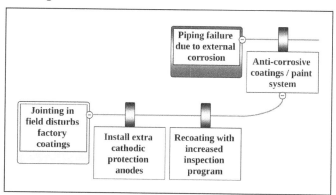

Figure 2-10. Example Placement of Degradation Control on Degradation Pathway

A degradation factor can apply to barriers on either side of the bow tie diagram. For clarity of visual appearance, often they flow from the left on the prevention side, and from the right on the consequence side, but they are the same in all other respects.

2.7.2   Degradation Controls: Concept and Location on Bow Tie

In the Swiss cheese analogy, degradation controls reduce the size of the holes in the main pathway barrier and provide greater confidence that the barrier will do its job effectively.

Degradation controls lie along degradation pathways into that barrier where they help defeat the degradation factor. Degradation controls should only appear on degradation pathways: they neither prevent a top event nor mitigate a consequence directly. Degradation controls often do not meet the barrier criteria of being effective, independent, and auditable although they will be stronger if they do meet these criteria. Similarly, active degradation controls may not contain all elements of detect-decide-act as this is only a requirement for barriers. They are frequently human and organizational factors concerned with the management of risk and barrier assurance.

During the creation of bow ties, measures may be identified that do not meet the criteria of a main pathway barrier. These measures are frequently degradation controls and it is important that they are captured on the bow tie. For example, the control 'recoating with increased inspection program', relates to degrading the coating system on piping (e.g., due to jointing in the field disturbing factory coatings) as shown in Figure 2-10.

Examples of degradation controls that rely on human and organizational factors include: engineering standards, contractor management, management of change systems, training, Job Safety Assessments, stop work authority, etc. In the example shown in Figure 2-11, the main pathway barrier is a passive barrier: steel containment envelope. A degradation factor for why the steel containment might fail could be 'inadequate review of design'. This degradation factor would be mitigated by two degradation controls: i) a formal design review checking application of standards, and ii) a HAZOP study to confirm integrity of design against feasible deviations. Neither of these has the capability to prevent the top event from occurring, but they both support and strengthen the passive barrier provided by the steel containment envelope.

It is a common error to place degradation controls onto bow tie main pathways. This causes confusion for two reasons: it loses the connectivity between which degradation controls are supporting which barrier, and it presents an incorrect visualization of too many barriers on the main pathway. This can give the impression of greater defenses in depth (that many barriers protect against that threat) than actually exist, and intuitively that the risk associated with the threat is adequately controlled, when in fact there are only two barriers.

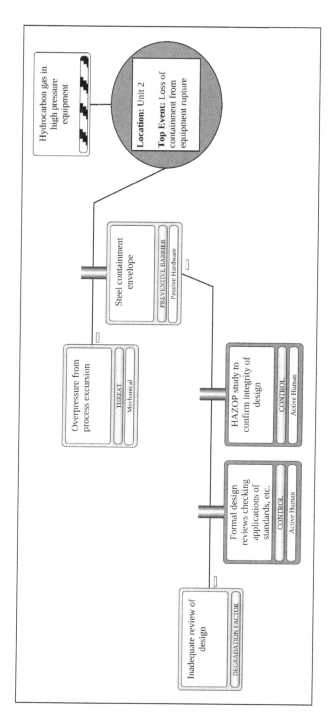

Figure 2-11. Example of Degradation Controls

A common example of a degradation control is a competence management system for the facility workforce. Competence management supports a main pathway barrier such as operator control of an exothermic reaction, but on its own it cannot prevent the top event 'loss of control of reactor'. Since competence management applies to most human actions, it may be appropriate to build out how competence management is achieved in detail only once and then address all other occurrences by reference. This is an example of where multi-level bow ties would make this approach efficient (see Section 4.2.2).

Multiple degradation factors can apply to a single barrier. It is common to have one or two degradation factors, but more than three can become complex. And, of course, degradation controls can themselves be degraded by their own degradation factors; this is an advanced topic beyond the needs of many bow tie users, and is discussed later in Section 4.2.2 under the topic of multi-level bow ties.

2.7.3 Use Degradation Factors and Degradation Controls Sparingly

One of the most important benefits of the bow tie method lies in its visual communication ability. Since the focus is on the more important barriers, the diagram complexity is favorably reduced by either not developing degradation pathways for less important barriers or, if these are developed for all barriers, then only displaying the less important ones when that additional level is important. Prior to developing degradation factors for every barrier, the value of this level of detail should be evaluated. Most bow tie software tools have several levels of display allowing the user to choose to hide some or all degradation factors from the general user, but show them for barrier owners or managers who need to know. Degradation factors are also included where there is a particular concern about a barrier or where historically the degradation factor has been a cause of barrier failure.

Degradation factors should not normally simply negate the barrier. It is generally not advisable to express degradation factors as the negation of the barrier. For example, if 'high level trip' is the main barrier, a degradation factor could be titled 'high level trip fails'. A better degradation factor might be 'level measurement device incorrectly calibrated' with degradation controls such as 'preventive maintenance for fluid level measuring instruments', 'instrument calibration', and then 'audit that the calibration is carried out'. The negation approach can lead to too many degradation factors and too general degradation controls making a diagram unnecessarily complex. A bow tie diagram should not attempt to capture every part of the safety and operational management systems. It is better to start with only limited degradation pathways for critical barriers where a need is demonstrated, with further degradation factors or degradation controls added only after the bow tie has been in use for a period. In some cases, degradation factors titled as a simple negation of the barrier may be appropriate if a more specific title is difficult to develop, but should only be used in a minority of cases for the reasons noted above.

Be clear about what it is that can cause the main barrier to fail. Degradation factors should consider the real reason for failure and not just that the barrier fails. Ask, "Why can the barrier fail?" Instead of stating 'procedure not performed', consider the reason – is it because 'contractors are not aware of local procedures'? 'Vehicles not to specification' might really be 'use of private vehicles'. The benefit of this approach is clearly that the degradation controls are then able to specifically address the real problem.

Repeating the same degradation factor on recurring barriers on a diagram may not add value. Degradation controls often describe the underlying, ongoing safe systems present at site (training, audits, inspections, back-ups, etc., are fundamental parts of the implementation of Risk Based Process Safety, see also Section 6.1). Once the first use of the degradation factor and its controls has appeared, subsequent equivalent barriers might not need to repeat the degradation model. Adding the text 'see degradation factor elsewhere' to the barrier description may be sufficient.

Sometimes degradation factors are defined on a more fundamental level and are part of the overall safety and operational management system. These issues, such as human error, communication failure, modifications to design and poor execution of the maintenance program are not specific to one barrier. The safety management system should have a verification and audit system for assuring the function of these items that are generally managed more efficiently in systems and tools other than a bow tie. Extensions to the standard bow tie to address deeper level degradation controls (i.e., degradation controls supporting degradation controls) are discussed in the multi-level bow tie approach presented in Section 4.2.2.

2.7.4 Degradation Factors and Degradation Control Examples

Well-worded and poorly-worded degradation factors and degradation controls are shown in Table 2-12 and Table 2-13.

Table 2-12. Well-Worded Degradation Factors and Degradation Controls Examples

Main Pathway Barrier	Degradation Factor	Degradation Control	Comment
Alarm and upwind mustering of staff	Wind direction unclear in congested plant or at night	Illuminated wind indicators on elevated equipment	This degradation factor highlights a specific circumstance in which the barrier may fail.

Table 2-12. Well-Worded Degradation Factors and Degradation Controls Examples, continued

Main Pathway Barrier	Degradation Factor	Degradation Control	Comment
Steel containment envelope	Equipment does not comply with process requirements	Formal design review against engineering standards. HAZOP study to confirm integrity of design. Asset integrity program to maintain the containment envelope.	This is a general degradation factor and it allows for the inclusion of multiple activities that need to be done to ensure proper design review.
Alarm and evacuation	New staff or visitors not trained in evacuation	Training for all new staff and visitors in evacuation.	This example highlights that training is usually a degradation control. (Note: evacuation alone is not a barrier as it misses the detect element required of an active barrier).

Table 2-13. Poorly-Worded Degradation Factors and Degradation Controls Examples

Main Pathway Barrier	Degradation Factor	Degradation Control	Comment
Pressure relief valve	No pressure relief valve	Design to include pressure relief valve	The degradation factor doesn't identify the real cause of the problem. 'Pressure relief valve removed for service' is an example of a credible problem, and allows a suitable safeguard, such as 'Back-up pressure relief valve'.

Table 2-13. Poorly-Worded Degradation Factors and Degradation Controls Examples, continued

Main Pathway Barrier	Degradation Factor	Degradation Control	Comment
Pressure relief valve	Pressure relief valve blocked	Periodic PRV bench testing	This degradation control is not correct because it does not act upon the degradation factor 'Pressure relief valve blocked'. The use of a negation for the degradation factor does not make clear that the cause of blockage is due to the closing of adjacent block valves.
Wearing a seatbelt	Forgetting to wear seatbelt	Airbag	This degradation control is also not correct, because it does not act upon the 'Forgetting to wear seatbelt'. It should be a main pathway barrier.

2.7.5 The Level of Detail Should Match the Goal and Audience of the Bow Tie

Any barrier can be described at different levels of detail. A generic formulation could be 'safe shutdown'. With a generic formulation such as this, there is usually only one barrier and the bow tie loses its informational value. A more detailed description could be ESD with manual activation showing the main pathway barrier and degradation controls supporting it – the degradation controls clearly show the detect-decide-act elements. A comparison is shown in Figure 2-12. The operator emergency training would have to include specific aspects related to when the ESD needs to be manually activated, not simply general emergency training. This is true for all training degradation controls – they must be specific to the barrier they are supporting.

2.8   CONCLUSIONS

This chapter has set out the basic components of a bow tie diagram and provided advice and some guidance for selecting and using those components. Good and poor examples are provided for all of the components.

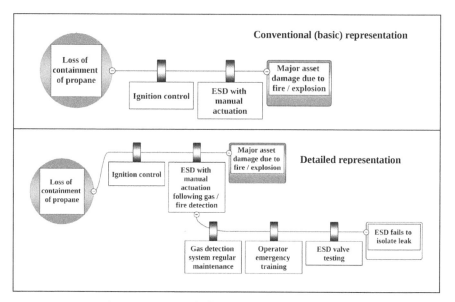

Figure 2-12. Level of Detail in a Bow Tie Diagram

The most important elements of a bow tie diagram, the hazard and top event, are reviewed, as they are the foundation of the analysis. Once specified correctly, the team can conduct a valuable analysis generating the rest of the bow tie elements. A barrier should satisfy all of three validity criteria: being effective, independent, and auditable (similar to LOPA IPLs) and if active, contain all properties of detect-decide-act. Using these criteria will generally limit the number of barriers to between one and five on a main pathway. Poorly constructed bow ties frequently show more than five barriers (and sometimes many more) giving both front-line operations and management the false perception they have more layers of protection in place than is truly the case. Measures such as training and inspection, which are sometimes found on bow tie main pathways, should appear instead as degradation controls on degradation pathways. Their role is to help prevent degradation of the main pathway barrier but do not themselves function as barriers.

Degradation factors can theoretically be developed for every barrier as these show the systems and hardware that support that barrier. However, this can lead to overly complex bow tie diagrams. Judgment is therefore required to select which barriers need communication to all readers. It is often better to have the capacity to hide some or all degradation factors and display them only for the relevant barrier owners or managers who need to know.

The quality of the bow tie as developed is discussed fully in the next chapter, with a checklist of key issues to verify in Section 3.3.

3

BOW TIE DEVELOPMENT

3.1 RATIONALE FOR BOW TIE DEVELOPMENT

Bow ties primarily provide a visually intuitive and readily understood model to communicate the causes and consequences of major hazards, and the barriers and condition of these barriers in a way that is easier for non-experts, workforce, engineers and managers to understand. The reasons and purposes for developing bow ties are described in depth in Section 1.4.

Bow tie development follows many of the normal PHA identification techniques. The Guidelines for Hazard Evaluation Procedures 3rd edition (CCPS, 2008a), for example, helps to identify the hazards and top events. Information from other PHA techniques can be used to support bow tie development. These include checklists, What-If, HAZID, HAZOP, FMEA, Fault and Event trees and other techniques. These capture key elements of bow ties: threats, consequences, barriers and degradation controls.

A characteristic of PHAs is their lengthy tabular format, which weakens their usefulness for communication. This is because PHAs may evaluate a wide range of scenarios, not just the much smaller number leading to MAEs. Bow ties are a useful means of facilitating this communication as they do focus on MAEs and they visually display the scenario with the threats, prevention and mitigation measures and potential consequences.

The next section outlines how to develop bow tie diagrams in a workshop setting. While workshops are a typical approach, simple bow ties or ones that are near duplicates of previously developed bow ties may be created by an experienced bow tie practitioner in consultation with domain experts. However, this loses the benefits of group thinking and buy-in by others.

3.2 BOW TIE WORKSHOP

The common starting point for bow tie development is with existing PHA studies, often HAZID or HAZOP. These will have identified a number of major accident events and many of the barriers / degradation controls deployed. If a LOPA exists it will help determine if the barriers are effective, independent, and auditable. Bow ties may also be developed after an incident to better understand existing barriers and degradation controls and to determine whether these are sufficient.

The process of creating a bow tie is most effectively accomplished using a bow tie workshop. It is important before the workshop to establish the scope and the context under which the bow ties will be produced. Table 3-1 provides a range of possible questions to consider.

Table 3-1. Questions to Consider for Bow Tie Workshops

Issues	Possible Answers
Who is the intended audience? (see also Section 5.2.1)	• Plant workforce (operations / maintenance); • Contractors; • Barrier and degradation control owners; • Local management and process safety specialists; • Unit designers (for bow ties during design stage); • Top management / board level; • Regulators; • The community; • Some combination of the above – but it is hard to satisfy all potential audiences with one analysis.
What is the study scope?	• Full range of events including smaller ones which may be important for safety, environmental or asset protection reasons; • Process safety only (e.g., loss of containment MAEs) or include events causing serious equipment damage (e.g., liquid overflow into compressor) or environmental impact only (e.g., smoke from flare); • Major health events; • Security events; • Whole plant or specified unit only (e.g., hydrofluoric acid alkylation unit); • Is there a need to capture barrier and degradation control metadata (e.g., responsible party, effectiveness, etc.); • Will risk ranking be conducted in the workshop?
What is the purpose of the bow tie?	• Training of staff on potential MAEs in the unit; • Major accident impacting enterprise risk; • Design review; • Periodic review of existing MAE defenses; • Assessing major Safety case; • Routine or abnormal operations; • Activity planning (turnaround, management of change (physical / organizational), decommissioning);

Table 3-1. Questions to Consider for Bow Tie Workshops, continued

Issues	Possible Answers
	• Analyze new technology; • Auditing; • Incident investigation.
Is there a prioritization of scenarios?	• High consequence; • High likelihood (e.g., potential repetition of prior accident event that has not been designed out, but not to address trivial events); • High risk; • Low level of automation (i.e., high dependence on humans); • Scenarios addressing concerns of regulator or local community.
Are there any specific human factor related issues or concerns?	• Is there an especially strong dependence on reliable human performance as a control against MAEs? (This may need the involvement of human factors specialists).
How many scenarios (top events) and will they be linked?	• Only MAEs identified in a prior PHA; • Larger set of events - brainstorm in workshop with no target number; • Will there be a consideration of relationships (chaining) between bow ties?
What method will be used for bow tie generation?	• Team-based brainstorming; • Specialist generated, updated by team; • Similar bow ties (e.g., from corporate headquarters, sister company, software vendor, or industry group).
Which rule set and terminology will be used in the study?	• Rule set and terminology suggested in this concept book; • Internal corporate guidance; • Software recommended set (contained in user manuals for commercial bow tie software).
Is there sufficient time and specialist team members available?	• Are a full range of skills available? • Limits may mean reducing the depth of treatment for all bow ties (e.g., limit number of threats or degradation factor developments), or limiting the total number of top events considered but giving these full treatment; • Prioritize the top events, with those that must be done and extras if time permits; • Request additional resources.

Table 3-1. Questions to Consider for Bow Tie Workshops, continued

Issues	Possible Answers
What consequences need to be assessed?	• Harm to people (staff and external populations); • Harm to specified environmental receptors; • Damage to assets / business interruption (financial); • Damage to reputation (non-financial); • Production, etc.
Which recording method will be used?	• Paper-based / sticky notes; • Word processor, spreadsheet, graphics package; • Specialist bow tie software – see Appendix A.

The workshop is a brainstorming session to assess the threats and consequences of loss of control of the hazards. It would confirm or improve risk treatment strategies so that risks are managed suitably and meet corporate requirements. Deciding on these issues will ensure that the bow ties developed are complete and match the objectives. The actual bow ties developed in the workshop may not be correct structurally and will need review after the workshop to avoid inconsistencies between different bow ties. It is not a good use of team time during the workshop to resolve details of bow tie structure and to cross-check other bow ties for consistency.

3.2.1 Bow Tie Workshop Pre-Work

A Terms of Reference, charter or PSO document (Purpose, Scope, Objectives) should be issued to the parties involved in the bow tie workshop. This document briefly describes the purpose and scope of the workshop, objectives, terminology, methodology, and ground rules that will be followed.

Prior to the workshop, generating draft bow ties based on past work or the previous hazard evaluation findings will make the workshop more efficient. Alternatively, the company may have a collection of previously approved bow ties to use as a starting point. This would provide a basic bow tie structure and allow the team to focus on specific differences, identify the correct barriers and systems in place, and quickly modify the bow ties to a final form.

However, care should be taken when starting with a previous bow tie. Although it can be efficient to start with a previous bow tie, it can hinder proper brainstorming. It is recommended to start by brainstorming on a clean sheet then switch to the previously developed bow tie to compare and discuss differences. If using previous bow ties, it is critical that the barriers be verified during the workshop; some barriers may not exist anymore due to differences in process, procedures or the equipment design. In addition, prior bow ties may not meet the rule set and criteria contained in this concept book and may need updating - for example, barriers that are actually degradation controls.

How many bow ties to develop? Determining the number of bow ties to develop is difficult. The UK HSE advises COMAH / Seveso sites in the UK to identify their major accident scenarios, then to select a 'representative set' and to perform a suitable analysis (e.g., bow ties) to provide an ALARP demonstration. Some early experience suggested that around ten bow ties might be sufficient for a major refinery or chemical plant. These would be a good starting point to gain experience in generating and using the bow ties in an operational environment and embedding their use within the operational management system. However, ten bow ties may not provide sufficient differentiation between the many major accident events that are possible and they might become too generic.

A better guide is that a major refinery or chemical plant activity might be expected to need between twenty to thirty (20-30) bow ties, but would be unlikely to greatly exceed that number. This number should be sufficient to address a wide range of processing, storage, and transportation issues, though always remembering that quality is more important than quantity. The objective is not simply to produce some target number of bow ties and then to file them. It is to understand and manage the barriers identified in an ongoing and effective manner. This is discussed more fully in Chapter 6.

If the organization has no experience of bow ties, then piloting perhaps two to three initially will likely prove more efficient as there is normally much iteration in both the design of the bow ties and how they are used in the organization. Experience shows that too many bow ties cause confusion to staff and detract from the objective of effective communication of risks and barrier management. More bow ties can be generated after experience is gained in their creation. Though it can be more important to ensure lessons from their regular use are incorporated as part of the risk management system.

Some companies have developed unit-specific bow ties. This generally generates a much larger number of individual bow ties, with significant duplication of major accident events (e.g., a pressure leak of flammable vapor on different units would be very similar). The benefit, however, is that accountabilities can be driven down to specific individuals rather than job functions and references to procedures and equipment failures, etc., can be specific to the unit. In addition, unit-specific hazards can be better addressed (e.g., medium pressure issues rather than addressing generic pressure hazards). Some hazards and related top events (e.g., hydrofluoric acid alkylation reactor, large LPG storage, or large inventory of highly toxic substances) would normally require a specific bow tie and will typically use multiple unit-specific barriers. A generic bow tie for handling pressurized flammable liquids and toxic substances would likely be insufficient for these higher hazard examples.

Documents needed. The following documents should be collected before the bow tie workshop so that the team can use them for reference.

- Process Flow Diagrams / clear description of the activity;
- The PHA and description of any process, production or 'environment' changes since the PHA was completed;
- Facility layout drawings;
- Previous bow ties from past work or PHA findings;
- Documents outlining the safety strategy or safety concept (e.g., EPA risk management plan or safety case document in other jurisdictions);
- Company risk matrix (as some bow tie software allows consequence classification by risk category);
- Best practice / corporate approved bow tie templates; and
- Industry guidelines.

For process systems, the following documents will assist the bow tie workshop, particularly if unit-specific bow ties are being produced or will need to be available outside of the workshop to check assumptions made in the workshop.

- Piping and Instrumentation Diagrams (P&ID);
- Previous PHA studies (HAZID, HAZOP, Risk Register, What if, and LOPA reports);
- Operations and maintenance procedures;
- Hazardous area classification drawings;
- Cause & effects diagram / shutdown key; and
- Hazards and effects register.

Information on previous incidents and failures will assist the understanding of what and how things can and have gone wrong. Examples include:

- Previous process safety accident, incidents, and near miss reports (including from related industries);
- Incident investigation recommendations;
- Relevant human factors studies or human error reports; and
- Audit findings.

3.2.2 Workshop Team

The success of the workshop heavily depends on the skill of the facilitator and the knowledge, experience, commitment and cooperation of the team. Team members are expected to give their undivided attention during the session, which is critical to the brainstorming process. Ground rules should include restricting use of computers, phones / mobile devices and other outside interferences during the session. This guidance is similar to that given for other team-based approaches (e.g. PHA studies).

The Facilitator. An experienced facilitator should lead the workshop. Alternatively, less experienced facilitators may conduct supervised sessions, observed by an experienced facilitator. The key role of the facilitator is to guide the process and to control and motivate the team. The facilitator should be knowledgeable in the bow tie methodology and have a working knowledge of other team-based risk assessment techniques such as HAZID, HAZOP, consequence analysis, LOPA and / or QRA. The facilitator should also have enough technical expertise to understand the process or operation being analyzed. Facilitators are often engineers (e.g., chemical or mechanical, for process industry studies). Strong communication and facilitation skills are needed to stimulate brainstorming, summarize the discussion and guide the team to consensus.

Facilitators should be familiar with other bow ties generated in the company to see how similar issues have been addressed before and to maintain a consistent approach.

The success of the workshop and the value of the output largely hinges on the facilitator's performance. Therefore, the contribution made by personal preparation, by the facilitator's preparation of the participants for their full engagement, and by attention to detail (e.g., workshop facilities, down to spare batteries and the right cables) should not be underestimated.

The Scribe. The bow tie scribe is typically someone who is not as experienced as the facilitator, but who has basic hazard evaluation experience. The scribe documents the discussions that take place during the workshop and records the first rough version of the bow ties as the team creates them. During the workshop, the scribe should generate a list of assumptions or questions posed by the bow tie team that are used to help simplify the bow ties and improve their clarity. Generally, this reduces the unnecessary build-out of the degradation factors. Some examples might be 'training and competency can be assumed in place for all procedures' or 'inspection includes calibration of any inspection devices.' The scribe will work with the facilitator to ensure that the recommendations are clearly worded. If proprietary software is used, then the dedicated scribe should be proficient in the use of the software and should, ideally, drive this software during the workshop.

The scribe has one of the hardest jobs in the room, trying to capture the final points from each discussion and record them accurately. His or her contribution should be recognized by the facilitator at the end of the session.

It is possible for the facilitator to act also as the scribe, as is sometimes the case with PHA workshops. While this reduces the study cost, it does generate extra workload for the facilitator and may slow the study speed. It also reduces the available thinking time for the facilitator, as this is often done in the pause while the scribe types the most recent entry. The tradeoff needs to be assessed by the person commissioning the study.

Team members. The composition and diversity of knowledge of the team is essential to a successful workshop. The team composition is similar to that for HAZOP or HAZID studies, but needs members who understand the full range of barriers and degradation controls deployed – this includes technical, human, and organizational barriers. The team should consist of Subject Matter Experts (SMEs) knowledgeable in the operations and maintenance of the facility and its controls; ideally, this should include a mix of disciplines (operations, maintenance, safety engineering, process control & instrumentation, procedures, risk management and process safety). Often a process operator from the plant will be involved as will a relevant control system specialist. If specific areas are to be built out, then additional expertise such as a human factors specialist or transportation manager may be needed. The team typically consists of five to eight individuals, but could be larger depending on the complexity of the process or scenario.

It is important to have a mix of technical specialists along with frontline personnel who are aware of day-to-day operations. Too large a team will make it difficult for the facilitator to control the proceedings. The core team should not be changed during the bow tie workshop and substitution of other team members during the study should be minimized. This will avoid the need to revisit already agreed definitions or approaches.

The team should be briefed on the study ground rules from the terms of reference or PSO document by the facilitator. They should also receive a short summary of the bow tie rules and terminology to be used for the study in case these are different from their past experience. If any team members are not familiar with the bow tie method, then a prior training session is recommended to ensure they understand the workshop process and the basic elements of a bow tie, use a common terminology, and understand how they can contribute their knowledge most effectively.

Often, not all team members will know one another and the facilitator should ensure all members introduce themselves and outline their experience. This helps the brainstorming process as members understand the background of the person commenting.

3.2.3 The Bow Tie Workshop

The workshop should be carried out as outlined in the terms of reference document issued prior to the workshop. Typical terms of reference should include objectives, scope, bow tie methodology, personnel required to attend the meeting, team responsibilities, schedule and deliverables, bow tie recipient (audience), distribution list and reference documents (HAZID, HAZOP, P&IDs, etc.). At the beginning of the session, the bow tie facilitator should provide a brief summary on the terms of reference, the standard bow tie method, terminology, rule sets, and differences between barriers and degradation controls – this is not the training referred to in the prior paragraph, but simply a restatement for people already familiar with the method.

If using software to record proceedings, it is advisable to regularly save the file and before leaving the venue, secure a backup copy.

Workshop session durations should be managed by the facilitator to balance speed of the study against the ability of the team members to concentrate and remain focused, which would otherwise detract from the quality of the result. This is very similar to workshop duration management in PHA studies.

During the workshop, the bow ties should either be drawn on a large sheet of paper visible to all or generated using software and displayed on a screen to aid in the process. Another highly effective blend of both is to use large scale paper or a whiteboard to quickly develop the diagram as the bow tie may become too complex. If the scribe has been following the paper graphic and developing in parallel the bow tie in software, then the group can then easily switch to the software version to continue its further development for completion up to adding the degradation factors, degradation controls and metadata.

Bow tie workshops are iterative and re-use bow tie fragments. Typically, there is repetition of prevention barriers between different bow ties and often almost complete duplication of mitigation barriers. Using the software to copy fragments from earlier discussions makes the process quicker and less prone to inconsistencies. A dual screen format is beneficial as extra information can be displayed (process information, brainstorm lists of threats, prior bow ties, etc.) while still retaining the draft bow tie image. The bow tie workshop is carried out using the steps shown in Figure 3-1. (In reality, the process is usually more iterative than this linear sequence suggests.)

The facilitator should lead the team in the following steps during the workshop:

1. *Identify the hazard and top event* or, if a draft bow tie was generated prior to the workshop, confirming the hazard and top event. The format should be similar to Figure 2-3 in Chapter 2.

2. *Identify consequences* through the brainstorming session. Identifying consequences before threats helps the team dimension the top event and later avoid threats that can only generate trivial top events and hence trivial consequences. This should then be reviewed against those generated from the PHA, if applicable. Confirm the consequences identified align at least to the study terms of reference. See Figure 2-4.

3. *Identify threats* through brainstorming that could lead to the top event for the given hazard and cross-check this against the PHA, if applicable. See Figure 2-5.

4. *Identify prevention barriers* once all threats and consequences have been identified. Systematically document all the existing prevention barriers in place and any potential new barriers for each threat identified. The identified barriers / degradation controls can be reviewed against those generated from the HAZID, HAZOP or LOPA if applicable. See Figure 2-6.

Some issues to consider when identifying barriers are:

*   Barrier name – clear and concise name that defines the purpose of barrier;

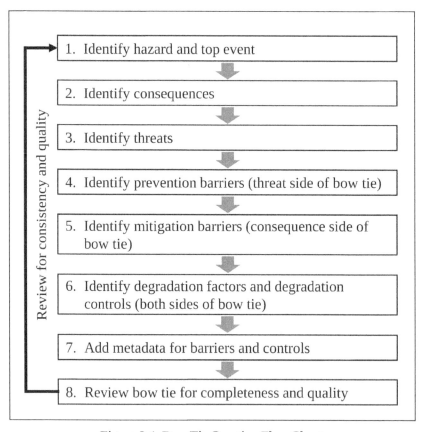

Figure 3-1. Bow Tie Creation Flow Chart

- Barrier validity – does the barrier conform to the validity criteria: effective, independent, and auditable? If it is an active barrier does it contain all elements of detect-decide-act, and if it is a prevention barrier, is it capable of terminating the event on its own (Section 2.6.1);

- Barrier type – define the type of barrier (Section 2.6.1);

- Hierarchy of control – use the type hierarchy or any custom list (Section 2.6.2);

- Barrier effectiveness – use an agreed rating system (some companies have created their own qualitative ranking system that teams can use to grade effectiveness consistently. This can be a numerical ranking or symbols such as +, o, – where + is stronger than an average barrier, o is average (performance similar to SIL-1), and – is poorer than average. Developing a rating system more complex than this is not recommended (Section 5.4.1);

- Number of barriers – verify what is reasonable (e.g., if the consequence is very serious then more or stronger barriers may be warranted (Section 2.6.3);

- Barrier criticality – if this is specified in the study terms of reference, this would review all barriers and assign a criticality ranking to these; and

- Document barrier metadata – performance standards, effectiveness ratings, critical tasks, responsible persons, etc.

5. *Identify mitigation barriers.* First document existing mitigation barriers in place and then define any potential new barriers for each consequence. The same considerations for prevention barriers listed above should be applied here as well. The bow tie should be similar to Figure 2-6 in Chapter 2.

6. *Identify degradation factors and degradation controls.* Some companies develop degradation pathways for all barriers, but most choose to develop these for more important or critical barriers, but not for all barriers. Identify degradation factors and degradation controls in place and then any potential new ones if the team judges that barrier strength needs to be improved.

7. *Add metadata for barriers and degradation controls.* This step captures knowledge of barrier owners, effectiveness, performance standard, current status, etc. using the knowledge available from the team (see the full list in Section 5.3.3).

8. *Review bow tie for completeness and quality.* This is to confirm that the terms of reference have been addressed and that a full range of threats and consequences has been included.

The flowchart shows a workshop iteration loop. It is unlikely that the bow tie will be completely correct as generated in the workshop. The workshop focus should be to use the time with the experts in the team to capture their actual experience of the barriers in use rather than whether the structure of the diagram is correct - this can be addressed in post workshop activities. The iteration loop then would be done offline and address the consistency and quality of the diagram. A checklist of quality issues is provided in Section 3.3.

The time required to create a bow tie depends on the complexity of the problem being addressed. Experienced teams can generate one to two bow ties per day, although the first bow tie can take considerably longer, just as the first P&ID in a HAZOP takes longer than remaining ones. This is due to the time to ensure the team understands all the concepts and terminology correctly. Also, results from the first bow tie can be used in subsequent bow ties without the detailed discussions required for the first.

If the bow tie workshop is running out of time, the facilitator should be prepared to defer any discussion and development that could be completed 'offline' without the full team. This may include the barrier type, performance standards and other links to critical tasks, safety critical elements (SCEs), management system references, etc. However, responsibilities and other barrier metadata should be discussed by the team.

3.3   POST-BOW TIE WORKSHOP ACTIVITIES AND QUALITY CHECKS

On completion of the workshop, the facilitator and scribe should meet to review the bow ties generated. These rarely meet all the rules suggested in this concept book and significant edits can be required. They should also consider other bow ties produced in the organization so that the style and terminology are consistent across multiple bow ties. The use of the preferred terminology from the Glossary is recommended.

Once the bow ties have been tidied up in the post-workshop activity, a formal quality review should be held. Typically, this can involve another experienced bow tie specialist who can provide an independent review of the bow ties produced and confirm that the approach matches other bow ties produced in the company.

The following prompts help to ensure the quality and consistency of review of bow tie diagrams.

Overall

1. Do the bow ties generated match the study terms of reference?

2. Have an adequate number of bow ties been developed to address the MAEs identified in prior PHA studies? A separate bow tie is not required for every MAE as some events may involve similar barriers and controls.

3. Did the workshop include the right mix of people; was an attendance sheet maintained; was sufficient time permitted to allow a thorough analysis; and has the quality review been carried out by an independent specialist?

4. Do all elements of the bow tie match the standard bow tie (hazard, top event, consequence, threat, barrier, degradation control – as covered in Chapter 2)?

5. Is the terminology used correct and consistent across all bow ties?

6. Does the bow tie contain any structural errors (e.g., degradation controls on a main pathway, or include ineffective barriers)?

7. Where multiple bow ties have been created, are they mutually consistent (e.g. similar top events call on similar barriers; and barrier names, degradation factors, and degradation control names are consistent with other appearances, etc.)?

8. Is the complexity and content of the bow tie suitable for the intended users?

Hazard and top event

9. Is the hazard clearly expressed with sufficient detail?

10. Is the top event a loss of control or loss of containment event, and not some type of consequence? Does it match the terms of reference for the bow tie workshop?

Threats and consequences

11. Are there too many or too few threats? For example, are multiple similar threats identified which result in nearly identical pathways? These would not add value but just more complexity to the diagram. Threats can be combined where the barriers deployed are the same (e.g., different but similar corrosion mechanisms).

12. Can all the threats lead directly to the top event and are they credible?

13. Are threats fully understood (e.g., unusual properties of materials or toxicities, necessary combinations to lead to the top event, etc.)?

14. If a threat is human error, should this better be treated as a degradation factor leading to impairment of a main pathway barrier? What is the main barrier that the human error defeats?

15. Do all threats lead to all identified consequences (via the top event)?

16. Do consequences cover the full range of significant outcomes?

17. Do all the consequences flow directly from the top event?

18. Where multiple bow ties are created, are the consequence categories the same (if appropriate)?

Barriers and degradation controls

19. Do all the main pathway barriers meet the required validity criteria?

    • Effective, independent, and auditable;

    • On the prevention side are they fully capable of terminating the event on their own; and

    • Do active barriers include all the aspects of 'detect-decide-act'?

20. Where barriers were defined sometime in the past, has engineering practice changed so that those barriers are no longer consistent with modern engineering practice (e.g., vents historically discharged locally to the atmosphere rather than as now collected and directed to a remote flare system)?

21. Are barriers included that cannot be measured, tested or monitored easily? This is part of the validity criteria for a barrier. If present, they should be redefined as degradation controls and moved to a degradation factor pathway.

22. Are multiple barriers included on a single pathway which have the potential for common mode failure?

23. Does the bow tie only list technical barriers and omit important human and organization ones? Note that an exception to this is design-stage bow ties that perhaps may only address technical controls, since the facility manning and management system has not yet been developed.

24. Have barriers been checked for less obvious dependencies (e.g., two barriers rely on the same operator, or if the detect-decide-act elements have not been fully stated then the omitted part conceals a dependency to another barrier)?

25. Have degradation controls been included on the main pathway rather than along a degradation pathway (e.g., many 'systems' belong as degradation controls not as barriers)?

26. Are barriers displayed in the appropriate time sequence of their operation?

27. Have barriers been cross-checked against legislation or recognized engineering practice to determine what barriers are expected to be in

place, but may not have been identified by the team? For example, safe separation distances based on facility siting (as per API 752 and 753) are now recognized good engineering practice and this barrier should be shown where appropriate on the bow tie if it has been done, or as a missing barrier if it has not.

28. Do mitigation barriers appear on the prevention side (e.g., ignition control can only apply after there has been a loss of containment top event) or vice-versa?

29. Have too many barriers been identified on any one pathway (usually due to placing degradation controls onto the main pathway or including trivial barriers)?

30. Are too few words used in barrier or degradation control names leading to possible confusion by users of the bow tie (e.g., 'certification' instead of 'crane load certification'; and 'procedure' instead of 'unit start-up procedure')? Frequently barriers and degradation controls are gathered together in lists separate from the original bow tie. Without this context, short names can be meaningless.

31. Do barriers and degradation controls include all company-required metadata, and at an adequate level? For example, do they identify who is responsible for each barrier operation, and avoid very generic terms such as 'Contractor' rather than clearly identifying the role as 'ABC Contractor Maintenance Manager')?

32. Are there too many barrier responsibilities lying with a single role? Distribution of responsibilities and resource management should be considered.

33. Has the combination of barrier types and their strength or effectiveness been analyzed to determine whether sufficient barriers are in place to prevent or mitigate an event?

34. Have past performance or incidents been considered in assigning effectiveness ratings?

35. Have critical barriers been defined along with associated supporting degradation controls to keep them at their intended functionality and links made to the company integrity management system?

36. Have trivial barriers been included (e.g., warning signage) that, if included at all, should be treated as degradation controls?

The use of this checklist should assist in delivering good quality bow ties, free of structural and other errors that might detract from their utility.

3.4   CONCLUSIONS

This chapter has reviewed the recommended process for creating bow ties. A workshop format is suggested with clear terms of reference agreed before the work begins. The team composition is similar to a PHA study. The facilitator and scribe require similar skills as per a PHA study to maintain a disciplined brainstorming environment and to record the discussions accurately. Team members require at least basic training in the method.

A methodical stepwise process is suggested for running the workshop to ensure all the elements of a standard bow tie are captured well and that any additional metadata is documented.

The number of bow ties required will vary, but for a large facility, twenty to thirty is often sufficient to capture the full range of top events leading to MAEs. For companies just starting to implement bow ties, a more modest objective might be to create two to three good bow ties initially rather than a larger number of poor ones. As experience in their creation and use increases, additional bow ties can be generated later.

A checklist of quality issues is provided that will assist facilitators and scribes verify that defects in the bow ties created are identified and rectified. However, a further independent check by a company specialist is recommended to ensure consistency with prior work and the company preferred approach.

4

ADDRESSING HUMAN FACTORS IN BOW TIE ANALYSIS

4.1 HUMAN AND ORGANIZATIONAL FACTORS FUNDAMENTALS

4.1.1 Introduction

Previous chapters have considered the origins and history of bow ties, the main elements of a standard bow tie, and how to create bow ties in a workshop environment. Most of these chapters have focused on technical barriers. The purpose of this chapter is to describe an approach to addressing human and organizational factors (HOF) that should be considered in developing and implementing bow ties. The book presents a multi-level approach to bow ties that works well for HOF and also other issues such as asset integrity. This chapter provides guidance on HOF both as potential threats and as barriers or degradation controls.

The chapter sets out a conceptual approach mainly for process safety or analysis of physical hazards. Within this context, it illustrates how a wide range of HOF can be addressed within a bow tie analysis to identify those that particularly need to be managed to ensure barrier systems are effective. Two approaches are presented: a conventional approach similar to earlier chapters, and a more advanced multi-level approach. The multi-level approach is novel and offers important potential benefits. The approach has been subject to thorough review and critical examination and is consistent with the white paper on human factors in barrier management issued by the Chartered Institute of Ergonomics and Human Factors (CIEHF, 2016).

4.1.2 Human and Organizational Factors - Conventional Approach

In earlier Guidelines, CCPS has defined Human Factors as 'a discipline concerned with designing machines, operations, and work environments to match human capabilities, limitations, and needs'. This has an ergonomics emphasis. IOGP provides a wider definition, including organizational aspects:

'Human Factors is the term used to describe the interaction of individuals with each other, with facilities and equipment, and with management systems. This interaction is influenced by both the working environment and the culture of people involved.'

This definition highlights the strong linkage between aspects of HOF and safety culture and is the approach favored by the joint CCPS and EI Committee.

In bow ties, HOF issues can appear in several places. Humans (including human failure – error or inaction) can be modeled a threat, but more often appear either i) as part(s) of an active prevention or mitigation barrier, ii) as a degradation factor, or iii) as part(s) of a degradation factor control. Humans can therefore form a barrier or a barrier element, provided the barrier meets the full validity criteria outlined in Section 2.6 – i.e., effective, independent and auditable. If it acts as a full prevention barrier the human activity must be capable of terminating the threat pathway on its own. Since a human barrier is always active, it must have all elements of 'detect-decide-act' present. Organizational factors or systems normally lie on degradation factor pathways.

Conventional thinking on human factors has often been concerned with errors. Modern thinking tries to integrate positive aspects of human behavior with negative aspects as shown in Figure 4-1. This figure shows expected and exemplary behaviors on the left side and errors and violations on the right side. Typically bow ties address human failures as degradation factors or as the holes in human related barriers. This figure is an update by Hudson for the EI of a well-known original figure by Reason. Human error as portrayed in this figure (slips, mistakes) and some violations (e.g., unintended actions) can be treated well in bow ties, but some other violations (e.g., reckless behavior) are not easily treated and workshop teams typically do not generate this type of violation in brainstorming.

The Bhopal incident is a good example of the difficulty of dealing with violations. Management there, while dealing with difficult economic conditions, shut down the refrigeration system and then sold the refrigerant. This disabled a barrier preventing exothermic reaction in the MIC tank if water contamination occurred. This might be categorized as an 'organizational optimizing' violation in the figure. Before the actual decision occurred, it is unlikely that any team of plant engineers and operators would have proposed that management would disable that barrier. They might have considered refrigeration system breakdown, but this is a lower likelihood event compared to complete removal. None of the degradation controls supporting the refrigeration system barrier against breakdown have any effect against deliberate shutdown and sale.

What should have been done was that an MOC should have been triggered by the change and that would have recognized that a safety barrier had been disabled. This should then have been followed by an assessment whether the remaining prevention barriers were sufficient – which they were not. Thus, some violations can be difficult to address in bow ties before the violation occurs, but might be captured by MOC reviews after the violation is discovered. It might be possible to have a degradation factor 'Unauthorized change to safety system' with 'MOC review" as a degradation control. However, this could apply to every barrier in a bow tie and it is better to leave this to the safety management system.

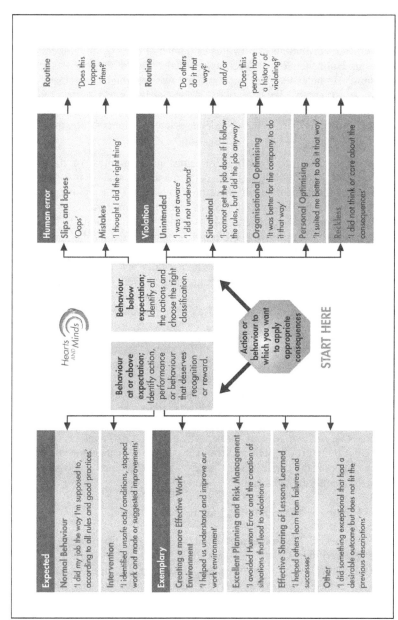

Figure 4-1. Positive and Negative Human Behavior Types

Source: EI Hearts & Minds, Managing Rule Breaking

4.1.3 Human and Organizational Factors - New Paradigm

Thinking of human contributions to bow ties as human failures greatly understates the true situation. Hollnagel (2014) describes a broader view of the role of humans in safety systems. He notes that historically safety was regarded as an absence of incidents or as an acceptable level of risk. Hollnagel terms this perspective as Safety-I: things go wrong due to technical, human and organizational causes – failures and malfunctions. In Safety-I risk assessments, humans are therefore viewed predominantly as a liability or hazard.

This perspective does not explain why things go right most of the time. Hollnagel notes that humans adjust their performance so that it matches the conditions. Safety management should therefore move from ensuring that 'as few things as possible go wrong' to ensuring that 'as many things as possible go right'. This perspective is termed Safety-II and relates to the system's ability to succeed under varying conditions. This perspective can be exemplified by the actions of pilots and air traffic controllers catching and trapping errors before they lead to unsafe conditions, including techniques and skills of 'Crew Resource Management' (Kanki et al, 2010). Although not so well established in the industrial process safety world, this approach of mindful sense-making and competent adaptive action is acknowledged as an important role of humans in process safety

A key advantage of a human barrier is that it can adapt to the situation in a manner that a technical barrier cannot. Simply indicating human failure as a threat or a cause of barrier failure does not recognize the positive contribution of the humans in the system. This is particularly true as system complexity increases and failure modes become harder to envisage. Hollnagel advocates a blend of thinking between Safety-I and Safety-II, keeping the better parts of conventional thinking, but recognizing the positive role of humans in adapting to novel or complex situations. This is typically achieved by focusing the management system components that support the barriers (see Chapter 6) on achieving better outcomes rather than solely identifying what could go wrong.

4.1.4 Human Failure as a Degradation Factor

The term 'human error' has sometimes been used as a main pathway threat. However, the required barriers can be very different based on the type of human error and the context in which it might occur. For example, referring to Figure 4-1, degradation controls against a slip (e.g., fatigue management) would be different to a mistake (e.g., refresher training); see also Table 4-1. The term 'human error' is usually too imprecise to be a good main pathway threat - it should appear as a specific degradation factor linking through the degradation pathway to a main pathway barrier. This is shown by example in Figure 4-2. This concept book recommends that human failure should not be used as a main pathway threat.

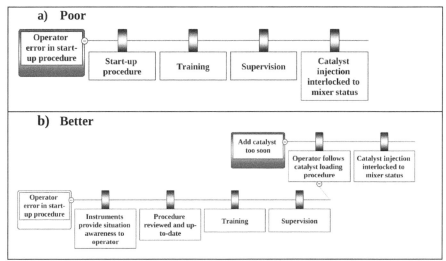

Note: The degradation factor pathway is bent around to fit the space

Figure 4-2. Poor and Better Treatment of Human Error in a Bow Tie

Human failure is situational; it occurs in a specific context that can include factors such as the design of the equipment being used, the working environment, the history of how the task has been done and what has been happening immediately prior to the activity. All these can influence how people perceive the situation and how they carry out the task.

A further problem associated with using human error as a main pathway threat is that it often leads to barriers that apparently match the threat but do not comply with barrier validity criteria. This is shown as an example of adding catalyst to an exothermic reactor before proper mixing is established. This can lead to a runaway reaction and an explosion. Part a) of Figure 4-2 shows a poor analysis driven by the human error threat, while part b) shows a better analysis where the human error is moved to a degradation factor. In part a), the threat which is human error leads the team to think immediately of procedures and training – which are not valid barriers. Part b) shows a better treatment. The human error is switched to a degradation factor and this acts on the 'operator follows catalyst loading procedure' for the reactor. The degradation factor shows that this includes all elements of detect-decide-act: degradation controls include instrumentation to detect the status of the reactor, a start-up procedure, training in the procedure and supervision, which together deliver the decide-act elements. In addition, the incorrect part a) sends a visual message of four barriers, when there are only two as is made clear in part b).

Figure 4-3 provides a further example of human error as a main pathway threat. This is a poor approach for several reasons: the threat (operator error) is too generic to permit meaningful barriers to be defined (i.e., what is the error?), and it leads the team to make errors in that several of the barriers fail to meet the validity criteria. 'Supervision' is too generic, as it does not clearly show the detect-decide-act elements. A start-up procedure is not a barrier as it is just a document.

A better representation of the same human error is shown in Figure 4-4. In this figure the threat is now specific (i.e., pressure deviation) and the barriers do meet the criteria required (as supervision has been moved off the main pathway and the start-up procedure is replaced by the key operator tasks defined in that procedure). Human error is dealt with appropriately in the degradation pathway and the error is more specific (i.e., 'start-up procedure not followed') acting on the main pathway barrier. Degradation controls on this pathway do not meet the criteria for a barrier, i.e., they are not capable on their own of terminating the accident sequence, but they do enhance the effectiveness of the main pathway barrier: 'operator tasks (unit start-up procedure)'. Supervision has been upgraded to 'active supervision during start-up' to indicate that it is expected that the supervisor is engaged in proactively reviewing conditions and activities, and not simply sitting in the office nearby.

Thus, a better way of treating human failure in bow tie risk management is as a degradation factor capable of defeating or degrading barriers. Presenting human failure in this way focuses effort on making the barriers robust by using degradation controls against the degradation factors (of which human failure is one).

4.2   STANDARD AND MULTI-LEVEL BOW TIE APPROACHES

This concept book offers two approaches for the treatment of human failures. There is the simple 'standard bow tie' (Section 4.2.1) using the approach already defined in earlier chapters but with some advice on where to place human failure threats. Another, more detailed approach, termed 'multi-level bow ties' (Section 4.2.2), allows a deeper treatment of human factors and can be generalized more easily across multiple bow ties where similar human failure potential exists. The decision as to which approach to use depends on the depth of analysis of human performance required and the skill sets available. Treatment of HOF as barriers or degradation controls is provided in Section 4.3.

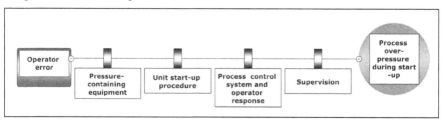

Figure 4-3. Poor Treatment of Human Error as a Threat

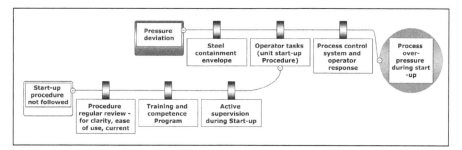

Figure 4-4. Better Treatment of Human Error as a Degradation Factor

4.2.1 Standard Bow Tie Approach

The standard approach is as presented earlier in this book, but with suggestions on how and where to include human error as a threat. It has the format as shown in Figure 2-1.

The standard bow tie approach will be appropriate for most MAEs. It provides an adequate level of detail for the workforce, engineering staff and local management. It does however lack details on deeper degradation controls that are particularly relevant for human factors and asset integrity where it is important to understand and manage these deeper degradation controls. However, its simpler format improves the ability to communicate its content.

4.2.2 'Multi-Level Bow Tie' Extension

The idea of a multi-level bow tie is proposed as a better means to explore human failure aspects in bow ties and to display the full range of degradation controls deployed. These are deeper level controls supporting standard bow tie degradation controls against their own degradation. While it does add some complexity, it builds on the existing bow tie and adds important extra information. It avoids duplication where similar human failure needs to be addressed in other bow ties as it can be reused without change, helping achieve consistency between bow ties. While the examples here relate to human factors, the multi-level approach can apply equally well to other topics where deeper level degradation controls are deployed – such as asset integrity.

This is a novel approach and users should not feel intimidated – the deeper analysis is intended for specialists and managers, not for all operational staff. Ultimately, the aim is to identify the full range of important HOF controls so that these can be understood by those who need to know. The UK CIEHF also recommends this multi-level approach when an organization wishes to explore human error pathways and the role of HOF controls in detail.

It is important to note that in the multi-level approach, the standard bow tie main pathways and degradation factor pathways *remain unchanged.* The extension shows how degradation controls in the standard bow tie can themselves be defeated or degraded and the additional degradation controls that might be needed; this extra level would be defined as extension level 1. Direct HOF degradation controls supporting a main pathway barrier appear in the standard bow tie, while more general level HOF degradation controls or regulatory requirements might appear in extension level 1. Degradation control examples at the standard bow tie level (i.e., those directly supporting a barrier) might include: procedures reviewed and up-to-date, training, and supervision; while extension level degradation controls (i.e., degradation controls supporting degradation controls) might include: operator recruitment psychometric screening, drug and alcohol testing, stop work authority, and senior management tours. A worked example is provided in Appendix C. In principle, extension levels could be worked out at progressively lower levels, though less benefit might be expected beyond level 1.

The concept of multi-level bow ties using existing software tools is shown in two formats: with standalone bow ties as shown in Figure 4-5 and   repeated   in Figure 4-6 with a format of cascading degradation controls. The upper part of Figure 4-5 is the standard bow tie with main pathway (only the threat side shown) and one degradation factor and its degradation controls. Extension level 1 duplicates the degradation pathway (shown here for Barrier 2 only), but built out again in the format of a standard bow tie. The hazard is the same, but the top event is now the failure of the barrier, not the original loss of control. Thus, the standard bow tie degradation pathway threat (Degradation Factor A) becomes the extension level 1 main pathway threat, and the new top event is the failure of Barrier 2.

This approach can be summarized as follows – multi-level bow ties start with a standard bow tie but make the failure of a main pathway barrier the top event of an extension level. They then examine how the degradation controls for the main pathway barrier can themselves be defeated and what needs to be in place to assure them.

This approach may be followed either manually (using paper, or graphics packages) or using commercial bow tie software to generate standalone bow ties that are easily reused elsewhere. However, it needs to be emphasized that the measures in the extension level are *all degradation controls*; barriers are only located on the main pathways of the standard bow tie. Figure 4-6 contains the same content as Figure 4-5 but now uses the ability of current bow tie software to add degradation factors and degradation controls supporting existing degradation controls. Bow tie software is not necessarily required, but it can make it easy to show or hide the extra levels depending on the needs of the audience. While simpler in format than Figure 4-5, this figure is less easy to re-use in other instances where similar barriers may fail and the full range of degradation controls is desired to be shown. Also, issued bow ties could become very complex where there are multiple main pathways and multiple level 1 extensions. An indication of this complexity is shown in the final figure of Appendix C for the overfilling of an atmospheric storage tank.

Extension level 2, if developed, would repeat the process, cascading to a further level of detail in an analogous manner to Figure 4-5 or Figure 4-6.

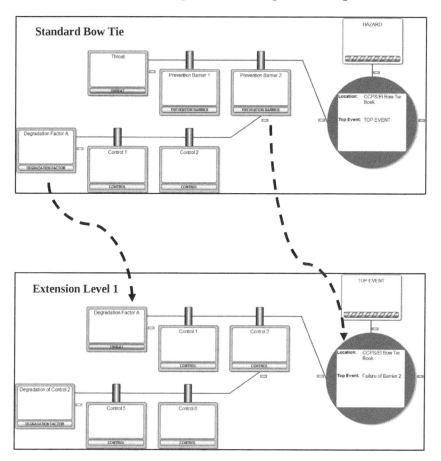

Figure 4-5. Concept of Multi-Level Bow Tie Approach (for Standard Bow Tie and Extension Level 1)

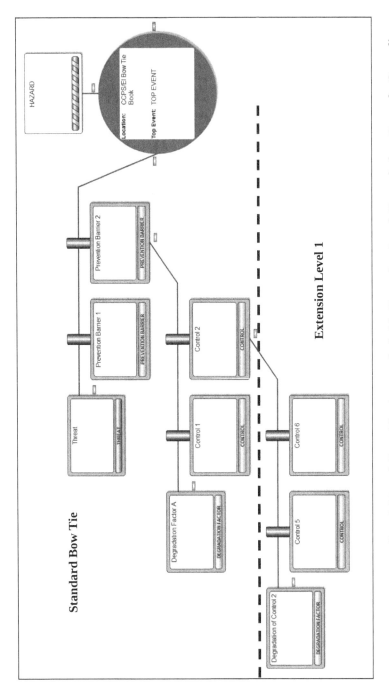

Figure 4-6. Concept of Multi-Level Bow Tie Approach (with Extension Level Degradation Controls Cascading Directly off the Standard Bow Tie)

Some examples of standard bow tie degradation controls and extension level 1 degradation controls are provided in Table 4-1.

In principle, this idea of standard bow ties with a level 1 extension applies equally well to technical controls, not just HOF controls. For example, if a preventive maintenance program is a degradation control in standard bow ties, then extension level 1 degradation controls might be maintenance team staffing, computerized maintenance program tools, and overdue maintenance reporting.

Examination of the extension level 1 degradation controls shows that these may apply to many human error threats. By breaking these out into extension level 1 bow ties as shown in Figure 4-5, then all these support systems are documented but with minimal duplication. This makes communication of the entire suite of bow ties much easier. All staff would be introduced to standard bow ties, and specialist and management staff to these plus the extension level bow ties.

Table 4-1. Examples of Degradation Controls: Standard Bow Tie vs Extension Level 1

HOF Degradation Type	Degradation Controls that might appear in a Standard Bow Tie	Degradation Controls that might appear in Extension Level 1
Slip	Fitness for duty	Drug & alcohol testing program Stress counselling Fatigue management Manpower and workload planning Medical assessment of fitness vs. foreseeable work tasks
Lapse	Supervision	Human factors in design of work system Supervisory skills training Active supervision (not administrative work) Fatigue management Teamwork programs encouraging communication and questions Safety culture program
Mistake	Training	Job design Human factors in design of work system Competence (general operator skills) Refresher training Training program periodic review

Note: HOF Degradation type – refer to Figure 4-1 for explanation.
Note: All degradation controls are part of the site safety management system and require active verification and review of gaps.

Another important benefit of the multi-level approach is that it increases awareness and visibility of the specific role that human and organizational elements are expected to play in the control of high risk hazards. For example, McLeod (2017) discusses the role of a 'STOP Culture' as a degradation control. Many companies try to achieve a culture where any member of the workforce who is concerned about safety is encouraged and expected to intervene to stop the work without blame or penalty. Recognizing the role of a STOP culture as a degradation control formalizes this aim. It makes clear to everyone who develops, authorizes, accesses or uses the bow tie – from senior management to the front-line workforce - the place that STOP culture has in the safe conduct of operations at the front-line. The same is true of many other administrative controls, including the use of engineering standards, contractor management, incentive schemes, etc. Representing them explicitly as degradation controls on bow tie diagrams makes their intended role in safety management clearer.

4.2.3 Comparison of Multi-Level and Generic Human Factors Bow Ties

Some companies have developed what have been referred to as "Generic human factors bow ties" with a top event of human error and typical human error threats such as fatigue, drug and alcohol abuse and task overload together with the safety management system programs deployed to reduce the likelihood of these threats. Such generic human factors bow ties typically do not comply with the rules for bow tie development (as set out in this book). For example, the nominated barriers are actually degradation controls, and the top event is not a true loss of control. In addition, the consequence side can be confusing as the barriers on this side depend on the actual operational situation (e.g., the so-called top event of human error will require different mitigations if the error type is a slip vs a lapse vs a mistake (see Figure 4-1)).

In some ways, the generic human factors bow ties approach is similar to the idea of the extension level bow ties in the multi-level approach, but it uses human error as the main pathway threat and this leads to structural problems as noted earlier. The multi-level approach showing extension level 1 for human factors issues linked directly to the standard bow tie is therefore recommended. This same extension can be applied to all other bow ties that need to show degradation controls against similar failures.

4.3   HUMAN AND ORGANIZATIONAL FACTORS AS A BARRIER OR DEGRADATION CONTROL

4.3.1 Barriers

HOF can be barriers meeting the full rigor of a barrier as defined in the Glossary and in Chapter 2. They must be able to completely terminate a threat sequence on their own; be effective, independent, and auditable; and since these are active, the

HOF (plus any associated hardware depending on the barrier type) must have all three elements of detect-decide-act. More often, HOF do not meet the full criteria for a barrier and instead they act as degradation controls supporting main pathway barriers.

Usually, 'administrative controls' will appear as degradation controls in bow ties. An example is the Lock-out Tag-out (LOTO) control. This might be thought of as a barrier, but the actual barrier would be the electrical breaker with the padlock. The LOTO control is a degradation control supporting the effectiveness of the breaker barrier.

An important rule is that HOF barriers will not be documents (i.e., a piece of paper does not make a facility safer), but entire systems that the front-line workforce are aware of and understand, with active steps and review. Some examples include (although these could also be degradation controls):

- Pro-active monitoring; i.e., operators with a realistic mental model of the entire process, actively seeking out information reflecting the state of a process or operation that is independent from alarm systems or other automated prompts. The full barrier title might be 'active monitoring of process state with shutdown option for deviation'.

- Work planning (this is the entire process, not just a sheet of paper). A barrier example might be 'work planning to avoid conflicting with simultaneous fueling operations'.

- Contractor management. A barrier example might be 'contractors use the site work permit system for the task'.

A HOF system that can be thought of as a barrier might be teams executing the hot work permit system (this is the entire process of the procedure, comprising the required training, nominated sign-offs, independent checking during the work, and regular post work document reviews). Against the threat of 'gas leak during hot work maintenance', the barrier might be 'maintenance team executes work permit requirements'. Thus, the barrier is the team working to the permit requirements, not the permit document itself.

HOF can also be part of an active barrier with detect-decide-act elements. A common example is: 'Alarm plus operator response within 15 minutes'. Here the alarm is the detect element, the operator is the decide element, and the specific response is the act element. Aspects that might degrade the operator element could be explored in a degradation factor with various degradation controls. These degradation controls (e.g., design of the alarm system, operator fatigue, or training and competence) would not meet the requirements for a full barrier, but they do reduce the likelihood of error and are important parts of the safety system. Such human degradation controls could become repetitive and would be better treated as an extension level 1 bow tie for that type of human performance impairment and listed only once. This is clearer and easier to communicate.

Usually systems and checking would appear in bow ties as controls on degradation pathways. Verification by a supervisor can be an exception and appear as a main pathway barrier where it acts as an independent barrier, not as a support, although systems should be put in place to ensure the supervisor is as independent as possible from the operator. An example might be a critical step in an exothermic reactor start-up where a supervisor must approve all conditions are appropriate before introducing the catalyst. Proactive operator monitoring can also be an important, and very widely used, main pathway prevention barrier (McLeod 2015). This is more than simply responding to alarms; it is a competent and alert operator actively seeking out other sources of information (e.g., CCTV system or radio field operator) to verify the state of the operation and systems. Effective proactive monitoring can overcome many possible weaknesses and lack of reliability in automated systems. In both cases the barrier should be described as what the supervisor or operator actually does, e.g., 'panel operator manages tank level by radio to field operator' rather than just 'operator monitoring'. The extra words make clear how the detect-decide-act elements work in practice.

A barrier should be effective, independent, and auditable. The latter two can be difficult to establish for HOF barriers. McLeod (2015) notes, regarding independence, that many apparently independent barriers do, in fact, have dependencies. For technical barriers, this can be relatively easy to spot. For a tank overfill example, a bow tie diagram might show three barriers related to tank level: target level alarm and operator response, high level alarm and operator response, and an independent high-high level operator-activated shut-off system. The first two use the same level detector – but if it fails, then both fail – so these together only represent one barrier. An independent shut off would have its own detect and act functionality, completely separate from the level detectors. In this scenario, however, both depend on the same operator. If this individual is absent, asleep, or pre-occupied, both barriers also fail (so the nominal three barriers become effectively just a single barrier). A better three-barrier model might be: 1) tank fill plan and proactive monitoring, 2) high level alarm and operator response, and 3) independent high-high automatic shut-off (with its own level detector and automatic shut-off). See Appendix C for further details.

Satisfying the criteria of independence in human factors terms can be challenging. Cross-checking is sometimes included to overcome the lack of independence, though this was reviewed post-Buncefield by the Process Safety Leadership Group (PSLG, 2009) and they note that this degradation control can be quite weak unless performed by a supervisor (e.g., due to operator reluctance to check a colleague's work in depth or due to a lack of skill). Supervisors also have other demands on their time, making the cross-checking control weak even when executed by a supervisor. The multi-level bow tie approach would make this clear.

4.3.2 Degradation Controls

As has been discussed, degradation controls appearing on barrier degradation pathways generally have a lower standard of validity than main pathway barriers. They support the effectiveness of a main pathway barrier, but do not terminate an incident themselves. Examples of degradation controls include:

- Competence (training and assessment);
- Inspection;
- Maintenance;
- Periodic audits;
- Safety culture programs; and
- Spot checks on operational discipline.

Degradation controls must still be subject to validation, however; otherwise, they may not contribute as expected to the main pathway barrier effectiveness.

4.3.3 Training and Competence

Training is a frequently misunderstood degradation control. Training is a specific activity, often in classrooms, but it may be on-the-job. It is harder to measure effectiveness of the latter type and ensure consistency between all staff – as is needed for effective work permit and LOTO operation, but it may be adequate for operational tasks. Competency is a broader concept and involves a wider collection of training, on the job work experience, and skills acquisition over time. While training and competency can be grouped within a degradation control (as they are dependent upon each other), they are different and need to be measured differently – training by testing and competency by a management assessment program. Training and competency in this context refer to degradation controls against degradation of specific main pathway barriers. This is not the total training program for an operator.

Task-specific training might support a main pathway active barrier (e.g., gas detection, operator decision, and push-button ESD). Such training must be linked to a specific safety critical task (the decision-making and the push button action) and both the task and training should be verified to be correct and relevant to the function of the barrier. In this regard, periodic drills help reinforce the required critical tasks and so provide the required validation because these may be checked in audits.

Training which is not specific to the barrier task is not a meaningful degradation control as there is no direct link to the required performance of the barrier. However, with such a link, it is an important element for the degradation control in supporting the barrier.

4.4   VALIDATING HUMAN PERFORMANCE IN BARRIERS AND DEGRADATION CONTROLS

In safety case regimes (e.g., Europe, Australia), it is a common requirement for facilities to nominate SCEs to define performance standards, and to validate these periodically (for details, see Energy Institute, 2010).   Design performance standards for hardware can be hard to measure during operations (e.g., blast wall strength) and a lesser acceptance criterion may be needed (e.g., no visible defects). These usually do not address human barriers.

Human and organizational systems can be considered as a barrier if they are auditable in a straightforward manner and meet the requirements for an active barrier (Table 2-9). Examples of the kind of tests that can be used to audit the status and validity of human barriers, or barrier elements might include (drawing on guidance on characteristics for a human Independent Protection Layer in LOPA (CCPS, 2001 and 2015; PSLG, 2009)):

- the process deviation and human action are independent of alarms, etc., already credited as part of the initiating event or other IPL;

- the action required by the operator is detectable and unambiguous;

- sufficient time exists to successfully complete the required action response in ten (10) minutes (higher error rate) or forty (40) minutes (lower error rate) as per LOPA (CCPS, 2001);

- operator decision-making does not require complex diagnostics;

- a procedure or troubleshooting guide exists including:
    o   a description of the hazard scenario
    o   process conditions that require a shutdown to bring the system to a safe state
    o   steps needed to correct or avoid the specified process deviation
    o   if the alert is from a safety system / alarm then full details of the operation of safety systems and specific operator actions to be taken in response to alarms
    o   verification by the operator that a safe state is achieved;

- normal workloads allow the operator to complete the actions required in the IPL;

- the operator can take the required actions under all reasonable conditions;

- the operator is trained in the specific procedure to follow;

- the operator can carry out the required action without personal endangerment; and

- human factors / ergonomic considerations have been applied in the design of the associated safety critical elements and related work systems.

As an example, consider 'maintenance team works to hot work permit requirements', as a barrier. The hot work permit itself would be a degradation control supporting this barrier. The maintenance team executing the hot work permit includes all elements for an active human barrier: detect-decide-act. Table 4-2 illustrates how this barrier could be assured and validated.

Table 4-2. Validating the Barrier: 'Maintenance Team Works to Hot Work Permit Requirements'

Permit Element	Requirement	Means to Validate
Work permit procedure	The procedure is complete and regularly reviewed. Required forms are clear, easy to fill out, and address expected hazards.	Auditor assessment
Nominated staff	The staff to issue, accept and be informed of the permit must be defined.	Auditor assessment
Training	All staff must be trained in what work requires a permit, the need to comply with the provisions, and nominated staff must have specific training for their roles.	Training records
Sign-offs	The permit form must clearly show places where sign-offs are required and by whom.	Document review and verification that the signers had the training and authority to sign
Job Safety Analysis	Before the work begins all members of the team must participate in a JSA, usually with all signing the JSA record.	Document review
Pre-work activities	The permit must specify any pre-work activities (isolation and purging, gas testing, electrical lock-out tag-out, sign posting, etc.) and these must be signed off.	Shift supervisor approval, Field operator approval and document review
During and post-work close-out	Operator patrols should periodically check on safety requirements and at the end of the work that all equipment has been returned to a safe state for start-up.	Document review and auditor assessment

Note: All the rows of this table are degradation controls supporting the specific requirements defined in the permit. Many of these may be logged instead as metadata for the barrier. This has the advantage of not making the bow tie too complex, but still capturing necessary validation activities.

Thus, a HOF barrier or degradation control may have many separate elements and separate validation will be necessary for each. To validate degradation controls, these generally need to have three properties (CIEHF website): 1) clear ownership, 2) traceability to elements of the management system, and 3) auditability.

4.5   QUANTIFYING HUMAN RELIABILITY IN BOW TIES

It is recommended in this book that bow ties, in the first instance, are not used for the purposes of quantification; thus there is no need to quantify human reliability for bow ties. It is however useful for bow tie facilitators to be familiar with human reliability, including quantification assessment, so that the risk of various threats or

barrier failures can be discussed and the need for more controls in degradation pathways can be assessed by the team to improve the barrier effectiveness. This type of assessment only indicates the order of magnitude, where an accurate assessment is not needed. CCPS (2015) Guidance on Independent Protection Layers provides some useful data tables for human response to an abnormal condition.

While quantification of bow ties is not encouraged in this book, it is noted that modern bow tie software tools do include the option to perform LOPAs on the bow ties developed as an additional activity.

4.6   CONCLUSIONS

This chapter has summarized how to address human and organizational factors in bow ties. Generally, using human failure as a main pathway threat leading directly to the top event is discouraged. This is because it tends to lead to bow tie structural errors with degradation controls appearing as barriers on main pathways and resulting in the false impression that more barriers exist than is actually the case. A better placement for human failure is as a degradation factor affecting a main pathway barrier. Most HOF controls therefore appear as degradation controls rather than barriers.

A novel multi-level approach for bow ties is presented that:

a)   allows progressively more detailed understanding of HOF degradation controls and how they themselves need to be assured, when such detail is needed;

b)   makes their role in controlling hazards clear and specific; and

c)   avoids duplication in multiple bow ties of similar lower level degradation controls.

The multi-level bow tie is not restricted to HOF: it can equally be applied for technical degradation controls, such as asset integrity, where there are a broader range of degradation controls deployed, but not shown on the standard bow tie.

The decision as to when to use the standard bow tie approach or the multi-level approach is up to the user. The standard approach works well with process staff and contractors as it is the simplest form of a bow tie and should lead to minimal confusion. The multi-level approach provides more detailed information on HOF degradation controls and would be relevant to HOF specialists, process safety professionals and managers. The multi-level approach at its highest level is a standard bow tie and thus can be used for communications to staff, but displaying the extension levels provides useful extra information on degradation controls.

5

PRIMARY USES OF BOW TIES

5.1   PRIMARY USE EXAMPLES

In this chapter, it is demonstrated that bow tie analysis provides a systematic structure and process to facilitate a qualitative understanding of major accident risks associated with a facility and the barriers and degradation controls deployed against those risks.   It can facilitate understanding by all staff and others by providing a graphic overview of scenarios for threat and consequence pathways, as well as allowing attachment of useful metadata (barrier owner, barrier effectiveness, barrier status, etc.).

Simplified graphics can provide for easier understanding of complex systems. Isometric layouts of process pipework, vessels and fittings may be technically accurate but can be too detailed to comprehend easily for a full process.   P&IDs, which do not show the geographical layout, are much better for understanding and these normally form the basis for PHA assessments.   Bow ties, in a similarly efficient, graphical manner, show complex arrangements of barriers and degradation controls protecting the system against threats and serious consequences.

The basic uses of bow ties covered in this chapter include:

- Linking bow ties to the risk management system and to highest risks as identified in the company risk matrix;
- Communicating major accident scenarios and all important barriers and degradation controls;
- Sharing barrier metadata;
- Monitoring the health of barriers and degradation controls;
- Identifying responsibilities for barriers and degradation controls;
- Allowing qualitative assessment of risk reduction measures; and
- Identifying safety and environmental critical information.

5.2   LINKING BOW TIES TO THE RISK MANAGEMENT SYSTEM

A comment sometimes heard about bow ties is that they are just a tool for communicating MAEs and barriers.   This greatly underestimates the potential value of bow ties in analyzing and assessing risk.

Bow ties can be used in the design phase to test the adequacy and relevance of barriers as initially specified, and to assist in deciding if additional barriers and degradation controls are required. Bow ties are primarily used in the operational phase to highlight key safety barriers, to support the assessment of their adequacy, to communicate this to all staff and contractors, and to provide a mechanism to continually monitor the effectiveness of those barriers.

5.2.1 Uses for Bow Ties – Design Verification

Bow ties are helpful in the design stage in assessing proposed barriers by their type and adequacy. They can:

- Provide a rigorous documentation of barriers preventing and mitigating MAEs. Bow ties provide a fuller list of barriers deployed beyond those normally found in HAZOP or PHA documents, thereby allowing exploration of degradation factors and their controls;

- Provide a structured approach to risk assessment for facilities that do not have P&IDs and so are less suitable for conventional HAZOP / PHA studies (e.g., mining, steel and other metal working industries). The team generates the barriers and degradation controls from their work experience without the need for P&IDs;

- During the design process or periodic review of the design, assist in determining whether an effective and sufficient number of barriers have been deployed for all major incident pathways; and

- Show whether diverse types of barriers are deployed (e.g., passive hardware, active hardware, active human, etc.) and in particular highlight pathways entirely dependent on humans.

5.2.2 Uses for Bow Ties – Communication and Management of Barriers and Degradation Controls

Bow ties can provide a range of benefits for communication and management of barriers and degradation controls. Examples include:

- They provide an easily understood format allowing stakeholders (workforce, management, regulators and the public) to understand the full range of MAEs and how these are controlled;

- Identify the barrier or degradation control owner who will take responsibility for each throughout the facility life;

- Highlight the important role of degradation controls, and;

- Provide a concise listing of all barriers and degradation controls to aid prioritization of inspection, maintenance and repair for engineering barriers and ongoing audits for administrative barriers. Note: this use highlights the importance of accurate descriptive labels, as stand-alone lists do not capture the context of the bow tie to make clear the barrier or control intent.

5.2.3 Uses for Bow Ties – Risk Management during Operations

Bow ties have multiple uses during the operational phase as well. These include:

- Risk management using bow ties to support answering questions such as:
  a. Are all the existing barriers functioning as intended? What is the status of the barrier against the design intent or performance standard?
  b. Are any barriers unavailable or deactivated on a temporary or long-term basis?
  c. Is it safe to continue operations or should the relevant activity be shut down until corrections are made?
  d. Are immediate mitigation measures required to strengthen barriers or are temporary additional barriers needed to allow continued operation?
  e. What is the prioritization for the longer-term actions to return barrier condition to the design intent or to meet the performance standard?

- Helping create a cumulative picture of risk management through the visualization of the number and types of barriers and degradation controls and their condition (e.g., visualization that multiple barriers on a single threat leg of the bow tie are not functioning as intended).

- Assisting in decision making for operations and activities when barriers are known to be degraded, or their status is unknown due to overdue maintenance or inspection backlog;

- Showing how individual barriers or degradation controls can address multiple MAEs, and help with change management to ensure that a change focused on one MAE does not adversely affect other MAEs; and

- Enhancing incident investigations, and improving sharing of lessons learned, by tracking all barrier failures (virtually all incidents involve some barrier failures).

5.2.4 Bow Ties and Links to Risk Based Process Safety

All organizational activities involve risk. Organizations manage risk by identifying it, analyzing it and then evaluating whether the risk should be reduced by risk treatments to satisfy their risk acceptance criteria. Throughout this process, they communicate and consult with stakeholders, monitor and review the risk, and verify that barriers and degradation controls intended to reduce the risk are functional, and that no further risk treatment is required.

At a high level, the risk management process can be described in three main steps (ISO 31000, 2009):

1. Establish the context (the nature of the operation, its design, and its environment);
2. Risk assessment (i.e., risk identification, risk analysis, and risk evaluation); and
3. Risk treatment.

The bow tie barrier approach addresses steps 2 and 3 of this process, i.e., the analysis and management of risks, and decision making about where additional risk reduction may be warranted. It provides support for conducting and documenting a risk assessment such that it is readily understandable to the target audience. And because they are qualitative and visual in nature, bow ties are accessible to a wider range of stakeholders than more quantitative techniques such as LOPA and QRA.

CCPS (2007) has published Guidelines for Risk Based Process Safety (RBPS), which is a comprehensive outline of the key elements in a modern risk management program. It has also developed a future view of the next generation of process safety management systems (called Vision 20/20, and available on the CCPS website). The EI has a similar framework (EI website, High Level Framework for Process Safety Management).

The RBPS management system incorporates four pillars with twenty core elements. The four pillars of RBPS are:

1. Commit to process safety;
2. Understand hazards and risks;
3. Manage risk; and
4. Learn from experience.

The following paragraphs illustrate how bow tie barrier analysis addresses all four pillars with the title for each of the 20 RBPS elements highlighted in **bold** below.

Pillar One: Commit to Process Safety.

- Bow tie diagrams help build a positive **process safety culture** as major accident risk management becomes more transparent to site management, staff and contractors;
- Bow tie diagrams build **process safety competency** – particularly major accident awareness during the operations phase;
- Bow tie diagrams foster **workforce involvement**; and
- Bow ties offer the realistic potential for **stakeholder outreach**.

Pillar Two: Understand Hazards and Risks.

- **Hazards identification and risk analysis** are the core subjects of bow tie diagrams; and
- The diagrams map how threats can result in loss of control of hazards, which can result in undesired consequences if prevention and mitigation barriers fail and form part of a site's **process knowledge/process safety information**

Pillar Three: Manage Risks.

- Bow tie barriers can be integrated into **operating procedures** and **safe working practices** explicitly documenting defenses;
- Bow ties document all the necessary barriers that are to be integrated in the **asset integrity and reliability processes**, in order to ensure they are available when required;
- Bow ties highlight the important role that degradation controls play in supporting barriers and in the overall management of process safety;
- Bow ties link accountability for each barrier to **contractors** and staff;
- Bow ties are an excellent **training** resource;
- Bow ties are an important source of information when assessing a **management of change** and determining **operational readiness;** and
- Bow ties can be used by **emergency management** personnel to ensure the mitigation barriers are in place and functioning for each scenario.

Pillar Four: Learn from Experience.

- Virtually every incident or near miss event means that some or all barriers did not function or barriers were missing – **incident investigation** can identify barrier weaknesses or where insufficient barriers are deployed;
- As proposed by IOGP (IOGP 556) **measurements and metrics**, i.e., KPIs should be aligned to barriers and their management;
- Bow tie diagrams underpin a barrier and degradation control **audit** program by highlighting all the barriers and controls relied on to protect against MAEs; and
- **Management review and continuous improvement** can be partly driven by focusing on barriers that have failed or found to be impaired in the preceding period and developing improvements to address these failures.

Bow ties complement quantitative and semi-quantitative risk assessment techniques (e.g., QRA and LOPA) which primarily address design decisions, though they serve a broader function and provide a greater focus on daily operations and decision-making.

5.3   COMMUNICATION OF MAJOR ACCIDENT SCENARIOS AND DEGRADATION CONTROLS

This section addresses the use of bow tie diagrams to communicate incident pathways leading to MAEs to the workforce, management and other stakeholders. This includes all important barriers intended to prevent or mitigate these events as well as degradation controls protecting against barrier degradation.

Hazard and risk analysis tools such as HAZID, HAZOP, LOPA and QRA can be complex. While necessary for their specific objectives, such methods can be poor for communicating MAE scenarios and associated barriers and degradation controls to the workforce and management. More detailed diagrammatic representations such as fault trees and event trees, while superior to text-based approaches, are still too complex for effective communications to non-experts. Some companies have developed large format artist-drawn diagrams showing MAEs and key barriers as scenarios. These can be effective for toolbox talks, though they are expensive to create and hard to update. The bow tie diagram conversely is straightforward and easy to produce, and, if using software, easy to update to accommodate changes or lessons learned.

5.3.1 Communication Needs of Different Audiences

It is important to recognize there are multiple audiences with different communication needs at various stages in the lifecycle of a facility or operation. The bow tie diagram can accommodate many of these needs, though other tools may also be useful. Table 5-1 provides a summary of different audiences and their typical risk communication needs.

Table 5-1. Risk Communication for Different Audiences

Audience	Risk Communication Needs and Bow Ties
Design team (capital projects)	The initial design team needs to focus on the technical components of barriers and highlight if important higher risk pathways have only human barriers. They also need to know where main pathway barriers rely on human performance in order to ensure that the design of work systems supporting that human performance fully comply with the principles of Human Factors Engineering. Some companies develop design stage FEED (Front End Engineering Design) bow ties that exclude human barriers. Any pathways with no listing of passive, active hardware or continuous hardware barriers will be empty and thus obvious to the design team of a reliance on human intervention, which the design may wish to address.

Table 5-1. Risk Communication for Different Audiences, continued

Audience	Risk Communication Needs and Bow Ties
Regulators	Regulators need an overall high-level understanding of the safety and environmental system functionality, similar to management. They may use bow ties to examine specific scenarios that have caused incidents elsewhere and to understand how identified deficiencies in barriers link to major accident events. The bow ties show the regulator the barriers and systems the company intends to have in place to manage the risk now and in the future. The Norwegian petroleum regulator (PSA) has an audit program based on reviewing barriers. There is also anecdotal information that the use of bow ties simplifies the approval of COMAH Safety Reports in the UK because it makes it easier for regulators to understand the risks.
Contractors	Contractors need to understand the risks in the facilities they are working on. The bow tie is an efficient means to communicate key major incident pathways and barriers, particularly those with which the contractors might interact. This could, for example, be recognizing the effect of their scaffolding in blocking optical gas detectors, electrical work removing power from key barriers, etc.
Management	All levels of management need to understand the major accident events and the main barriers and degradation controls at their facilities. Managers assigned to new facilities need to quickly understand all key safety information as well as the business issues. HAZOP worksheets are poor for this, whereas the bow tie is much more effective. Managers need to know what the main risks are at their facility and how they are safeguarded. Bow ties can provide information and direction to allow managers to ask probing questions about the health of particular barriers.
Local Community	Facilities sometimes need to communicate safety and environmental information to the public or their representatives in a manner that is understandable and effective. Company representatives can then address public concerns about certain pathways, show how multiple barriers are deployed to protect against such an event, demonstrating that the company is in control of the hazards, and that risks are minimized. Bow ties move the discussion beyond platitudes that 'everything is under control' to demonstrate that specific controls are in place for identified MAEs.

Table 5-1. Risk Communication for Different Audiences, continued

Audience	Risk Communication Needs and Bow Ties
Top Management / Board Level	A special class of user is the top management of an organization, head of operations, the CEO and the Board. After the Texas City accident in 2005, it was recognized that top management has a responsibility to understand process safety and to be aware of how potential major accident events are prevented. The bow tie process is ideal for this purpose. It tracks MAEs and demonstrates that there are sufficient safety barriers to prevent and mitigate all pathways. This is the assurance that top management needs – not so much the detail, but that all the pathways are under good control. Their ongoing responsibility is to provide the leadership, commitment and resources to ensure the barrier management process is working well so that all barriers are meeting their expected performance requirements, and managing the overall risk to the enterprise.
Workforce	Bow ties clarify the role of the workforce in operation and maintenance of safety critical elements and accountabilities for safety critical tasks. Metadata, included with the diagram, can show important additional information such as barrier owner, barrier effectiveness and reliability, and any status information (e.g., currently degraded pending the next turnaround). Color-coding of the barrier can be used to show which operating group is responsible for each barrier – operations, maintenance, engineering, etc. Bow ties used for operations do not include all the elements of the facility safety management system. Other mechanisms will be required to communicate this, such as training courses, toolbox talks, posters, etc.

5.3.2   Presentation of Bow Tie, Barrier and Degradation Control Information to Different Audiences

Different levels in an organization have different information needs. This can be addressed by developing bow ties with multiple levels of detail. For instance, a manager may only require a high-level bow tie showing the barriers at a facility level, while frontline personnel need a detailed bow tie containing information on barriers specific to a specific location, task or operation.

Bow ties can be drawn manually, but as for most complex tasks, specialist software can make life easier, especially where the same information needs to be presented in different levels of detail for different audiences. Most commercial bow tie software tools offer this functionality. CIEHF has highlighted that the use of software does not guarantee the analysis is correct and users must be careful not to be constrained in their thinking. Bow ties are frequently used for multiple purposes, and for different audiences as shown in Table 5-1. The structure of bow ties is such that safety barriers can contain critical risk information. The following indicate some different formats that bow ties can be presented in, for different applications.

- *Introductory Level.* This level might only display the hazard, top event, threats and consequences (i.e., the combination of Figure 2-4 and Figure 2-5). This compact format identifies the relevant MAEs for that hazard and the threats that need to be managed. If there are several bow ties, this compact format allows readers to quickly identify the bow ties they wish to examine in more detail.

- *Standard Bow Tie Level (without degradation factors).* This level displays all the information in the introductory level, but now adds prevention and mitigation barriers and all degradation factors and degradation controls (as shown in Figure 4-5). This extra level of detail is the most common format and would be used to lead discussions with a wide range of workforce and managers as to MAEs and all the barriers and degradation controls.

- *Standard Bow Tie Level (with degradation factors).* The level of display of degradation factor pathways would be decided by the needs of the audience. This extra information is clearly required by barrier owners and managers as the degradation controls are the measures that help to maintain the barriers at their expected level of performance. But they do add to diagram complexity and can defeat a key objective of easily understood communication.

- *Enhanced Standard Bow Tie Level (with metadata).* This is the same as above but with relevant metadata displayed beneath each barrier. This could be any of the information listed in Section 5.3.3. Examples might be barrier owner, barrier type, barrier functionality at last test, date of last test, and barrier equipment documentation. This would be particularly relevant to the barrier owners. Some different metadata could be criticality, acceptance criteria, and measurement type that might be more relevant to operations staff.

- *Multi-level Bow Tie Format.* This format was described in Section 4.2.2. This format shows lower level degradation controls supporting standard bow tie degradation controls (i.e., controls supporting controls). As discussed in that chapter, even though they are not shown in standard bow ties, these are important controls. It is important for managers to understand the role that these deeper degradation controls, such as safety culture, play in managing major accident risk, so they keep them active and functional. However, not all staff need to see this level of build out.

It makes the diagram more complex (as shown in the example in Appendix C) and it could defeat the communication objective.

An advantage of bow tie software is that most of the software packages support different levels of display, while storing the full details. Thus, the display can match the audience needs with minimal additional complexity, but if a question arises during the presentation then extra levels can be displayed.

5.3.3 Documentation of Metadata

The term metadata is normally defined as "information about other information". In this context, the base information is the barrier name and metadata is the collection of other data relating to the barrier. Table 5-2 provides a typical list of metadata; however, this list is not exhaustive and commonly used bow tie software can support a wide range of metadata fields. The barrier owners will need the most information for their barriers. Alternatively, auditors will wish to see all barriers that are measured by audit – to ensure they do not omit that activity during a plant audit.

Although there is no limit to the amount of metadata that can be displayed, excessive detail can create complexity and confusion, and may hide the message of MAE control, thereby defeating the aim of communication. As such, Table 5-2 lists the primary or most common types of metadata. Cascading levels of detail can suit specific purposes, and the level of detail displayed in bow tie diagrams should depend on the audience and the purpose and relevance of the bow tie information, so a specialized audience may need the extra metadata as described in Table 5-3. Barriers can also be color-coded to indicate various assigned properties or operational responsibility; other options include font type and size, attached notes, etc.

Effectiveness / Inherent Strength. While every barrier should fulfill the criteria of being effective, independent, and auditable, some barriers are better than others. When considering how well the risk is being controlled, it is important to not only look at the number of barriers but also how well each barrier performs. For instance, two 'high performance' barriers (SIL-2 rated or equivalent) can provide better risk control than three or four barriers that perform less well (SIL-1 or not rated).

To be effective, a barrier has to be 'big enough', 'strong enough' and react 'fast enough' to stop the threat leading to the top event or to mitigate the consequence.

Effectiveness options to characterize the effectiveness of a barrier include:

a) qualitative rating (e.g., ++, +, 0, -, - -)

b) quantitative estimates (e.g., probability of failure on demand) or SIL rating,

Table 5-2. Primary Types of Barrier Metadata

Data Type	Description
Barrier type	Passive Hardware, Active Hardware, Active Hardware + Human, Active Human, Continuous Hardware.
Barrier description	Additional description of barrier (e.g., procedure number, training course reference, equipment / components of the barrier, equipment detail drawings, and manufacturer information).
Barrier owner	Job title of owner – Maintenance Manager, HSE Supervisor, etc. Generally taken as the person accountable for the correct operation of the barrier. For example, the plant operations manager would be accountable for ensuring an ESD valve functions correctly even though maintenance maintains it, whereas a relief valve would be owned by the maintenance manager as no operational activity is required.
Performance standards	Typically, specification of Functionality, Availability, Reliability, Survivability, Dependency and Compatibility (e.g., ESD valve reliability, closure time and leak rate). Performance standards (for Safety Critical Elements or systems) are a regulatory requirement in several countries (e.g., UK initially, Australia and all EU countries later).
Effectiveness / Inherent Strength	An assessment of the 'as designed' performance of the barrier. Reliability / Availability and Adequacy are sub-elements to assist the understanding of the effectiveness of a barrier
Criticality	In principle, all barriers are important, but if criticality were used then metadata would identify that the barrier is on a safety critical register or indicate its ranking in the Asset Integrity program.

c) By hierarchy of controls, the type of barrier tends to indicate its effectiveness / strength, i.e., 'passive hardware' being the strongest, followed by 'active hardware', then 'active hardware + human' (with only human element) and finally 'active human' (with all elements being human).

There are two sub-elements to effectiveness / strength which can assist the understanding of the effectiveness of a barrier:

Reliability/Availability. Reliability refers to the chance that the barrier will perform when called upon. Availability refers to whether it is in service when the demand occurs. There are many reasons why a barrier might fail. The barrier might not be maintained or be turned off and hence not operate when called upon

Table 5-3. Supplementary Barrier Metadata

Data Type	Description
Condition (status / degraded)	Indicates whether the barrier is degraded from its intended design state or acceptance criteria, see Section 5.4.4.
Acceptance criteria	Criteria to determine if the barrier is degraded (e.g., at least 5000 gpm firewater flow, at 120 psi at the most remote location, within 3 minutes of actuation). This is not necessarily the same as the performance standard where this can be difficult to measure in operation.
Measurement type	How to determine if the barrier is degraded (inspection, testing, observation, and audit).
Support frequency	Frequency of preventive maintenance, retraining, exercises, etc.
Maintenance type	Type of preventive maintenance required, spare parts that must be in inventory.
Corrective actions	When degraded, what remedial activities are required for this barrier (normally short to medium term, and could include managing barrier-related information).
Critical tasks or critical elements	What tasks need to be conducted to maintain the barrier? For example, via a hyper-link to a specific operating / maintenance procedure
Documentation	What documentation needs to be linked to this barrier? This could include manufacturer, corporate, industry or regulatory information.

to work. A dike (bund) to prevent tank spills from reaching a sensitive environmental receptor is a passive barrier that should be highly reliable (quantitatively, having a low probability of failure on demand). The failure mode might be structural failure of the dike. However, many dikes include drain valves to remove collected rainwater and there is a chance of failure if poor valve management allows it to be left open. A more reliable option to increase the reliability of a dike would be to use a sump pump, with no breach of the dike.

Adequacy. Even if a mitigation barrier performs when called upon, it is not necessarily completely effective. Adequacy is the barrier's ability to control the threat or consequence when performing correctly. For example, a water spray system protecting against toxic ammonia vapors is never 100% effective; often 50% might be expected and possibly close to 0% if the wind is blowing in another direction.

A barrier that is present multiple times in one bow tie may have variable effectiveness across the bow tie. For example, the effectiveness of the barrier 'proactive operator monitoring' would be affected by the monitoring information available, how easy this is to access, distractions and competing tasks, and the time available for a decision.

Criticality. Another useful type of metadata is the criticality of a barrier. It can be argued that all barriers on a bow tie diagram are critical, so a category of criticality is superfluous. However, many organizations do have criticality management systems and the bow tie can fit into such a system. Different levels of criticality can be assigned to a barrier, such as high, medium or low criticality. Assigning a criticality level to each barrier shows that this aspect has been assessed. The purpose of this categorization is to indicate which barriers need more focus and ongoing monitoring, maintenance and immediate rectification when required. When a critical barrier fails, it can be assumed that the risks associated with the threat under consideration are greater so the chance of an undesired event occurring increases significantly, warranting the additional focus and attention on that barrier. Pressure relief valves might be determined to be critical as the last line of defense and in most jurisdictions having a functioning pressure relief valve is a legal requirement. See also later discussion on barrier criticality in Section 5.5.1.

5.4 USE OF BOW TIES IN DESIGN AND OPERATIONS

Bow tie diagrams show all the important pathways to major accident events and the intervening barriers and degradation controls. An important use of bow ties is the assessment of adequacy of the barriers and supporting degradation controls.

Being clear about the required effectiveness of barriers will help bring focus upon how barriers and their degradation controls need to be managed to ensure they provide the level of protection expected and intended.

5.4.1 Determining the Sufficiency of Barriers in Design

A high risk pathway is one on the prevention side with a high frequency threat. High risk pathways generally need greater risk reduction than low risk pathways, but determining the required number and effectiveness of barriers on each threat and consequence pathway is difficult. There have been some attempts at counting barriers with each barrier assigned a score considering reliability and adequacy to determine sufficiency, but this becomes an imprecise LOPA and is not recommended. A rule-based quantitative approach also may act against the pursuit of the ALARP principle – if it is easy and low cost to implement a risk reduction measure then it should be done regardless of some barrier count. One more barrier may be deemed superfluous in the 'count' but could be an obvious 'yes' for inclusion if it reduces risk significantly with low time, money and effort to implement and maintain.

In assessing the number of barriers, only the main prevention and consequence pathways should be considered, not the degradation pathways. The degradation factor pathway demonstrates how reliability is achieved for the main pathway barrier, but degradation controls do not increase the main pathway barrier count.

During the design stage the normal design hazard management process, including PHA studies (i.e., HAZID, HAZOP and LOPA), support the design of the system that determines number and types of barriers deployed. LOPA provides a detailed rule set and semi-quantitative methodology (order of magnitude) to determine the required risk reduction through barriers, but does not cover all the barriers and degradation controls shown on a bow tie. Bow ties can be used early in the design stage to assist in deciding whether sufficient barriers are in place. This is a qualitative decision, but the diagram makes explicit the barriers in place and supporting degradation controls, and these can be compared to other pathways and other bow ties to ensure a consistent approach to managing risk. More commonly, bow ties may be a final design stage product supplied as an input to the operational phase.

As discussed previously, some companies only show hardware barriers in design-stage bow ties, as the operational management system is not yet in place. Pathways with no barriers in this approach indicate that only active human barriers will be used and these may be good candidates for adding hardware barriers.

5.4.2   Initial use of Bow Ties in the Operational Phase

Although an operations representative will normally have been present when the bow tie was created, a wider team of operations and maintenance personnel need to familiarize themselves with the bow ties and in particular the barriers and degradation controls.

The initial steps for the operations and maintenance team to adopt their bow ties are as follows:

- Understand which bow ties are available and for which hazards and top events;

- If 'corporate' bow tie templates have been used, then a mini bow tie workshop may be required to make the bow tie specific to the system where it will be used;

- Check that the barriers from the particular system conform to the effective, independent, and auditable criteria. If the barriers are active, all three elements of detect-decide-act should be clear, check whether prevention can terminate an incident sequence on its own.

    o   This will help a wider team of operations and maintenance personnel understand which barriers are present in their system and the rules for barriers.

- Review the degradation factors and degradation controls.
  - o This will likely be a continual improvement process as the bow ties are in use over a period of time. It is better to start using the bow ties rather than wait until the degradation factors and degradation controls are 'perfect'.
- Assign barrier owners and if possible degradation control owners.
  - o Owners help ensure not only the forward reliable operation of these measures, but also that the assumptions underpinning them are not lost in the transition from design into operations.
- Agree to a schedule to review the risks using the bow ties.
  - o The details of these review sessions are in the following section, but additionally a schedule can be created for certain barriers and degradation controls that are subject to detailed review. The schedule can be arranged so that (ideally) over a six- or twelve-month period, all the barriers are reviewed. These reviews should include a review of the barrier effectiveness.

5.4.3 Operational Phase Barrier and Degradation Control Reviews

The aim of ongoing bow tie-based risk review meetings should not be to 'rubber stamp' the continued operation, but rather to be truly inquisitive, to challenge each barrier with a view to discovering the 'holes in the cheese' before devising improvements. A multi-discipline team of operations, maintenance and engineers should support the risk review meeting and have engagement and support from management. Within a single risk review meeting there is typically insufficient time to fully explore in depth the current condition of all barriers and associated degradations controls. All risk review meetings should cover known significant barrier or degradation control condition concerns together with an in-depth review of specific barriers and associated degradation controls conducted on a rotation basis. In this way, after say twelve months of meetings (e.g., as an addition to a regular unit management review meeting), all the barriers have been explored in depth.

There are five key questions when reviewing barriers and degradation controls using bow ties:

1. Consider the threats - are there any changes in the context or demand under which the barriers operate (e.g., new threats, or changes in the throughput or environment)?
2. What is the current barrier condition?
   - o Are all the existing barriers functioning as intended? What is the status of the barrier against the design intent or performance standard?
   - o Are any barriers unavailable or deactivated on a temporary or long-term basis?

3. Is it safe to continue operations or should the operation be shut down?

4. Are immediate measures required to strengthen barriers or should temporary additional barriers be added to allow continued operation?

5. How are the longer-term actions being prioritized in order to restore barrier condition back to the design intent, or to meet the performance standard?

By exception, additional barriers might be proposed.

It could be argued that the facility's existing process safety management program is sufficient to manage barriers. However, even good operators with effective process safety management programs can have barrier failures that may result in major accidents. Chapter 6 addresses how a barrier management program might be implemented.

5.4.4 Representing Barrier Condition on a Bow Tie

The effectiveness of a barrier is a function of both its condition and its inherent strength. The condition of a barrier is different from its inherent strength. Inherent strength is the effectiveness of the barrier in its 'as designed' condition or condition when the barrier fully meets its acceptance criteria. Inherent strength can be thought of as the number of holes in a slice of Swiss cheese, whereas condition would be the size of these holes. Condition is the state of the barrier at a specific time. There is currently no consistent agreed way to display barrier condition on a bow tie diagram: it can be as color codes or as text beneath the barrier or even as special symbols. The representation may depend on the tools available in the software used. Note that human factors specialists suggest that only one color system should be applied on any bow tie - whether this is for responsible party, barrier effectiveness, or barrier condition. Multiple color systems can be confusing to users. Software tools allow patterns to be used instead of colors, but this can also be confusing to users.

With that caveat, one system used successfully by a major oil and gas company is shown in Table 5-4. The color-coding can be applied as shown in the table, and different software tools display these in different manners (e.g., as a note under the barrier symbol or by coloring the barrier bar). Some other companies use the sequence of symbols +, 0, -, placed beside the barrier to indicate barrier condition, though the same symbols are also used for barrier effectiveness (inherent) strength, which is a different characteristic that addresses effectiveness.

Figure 5-1 illustrates the implication of barrier degradation on increasing risk and the means to return the system to the original or to a desired risk level (e.g., risk reduced to ALARP). In the 'design basis' three barriers are shown to achieve an acceptable risk target. For 'during operations', barrier 1 is shown as failed and barrier 3 is in a degraded state (e.g., overdue maintenance due to parts unavailability). The risk target is now not being met as two of the three barriers

Table 5-4. Suggested Formats to Display Barrier Condition

Condition (simple)	Condition (detailed)	Color code
Effective	In place, available and effective	Green
Partially effective	In place and available, but operating below its intended functionality	Yellow
Not effective	Not in place, not available	Red
No data	No operational information is currently available	White
Deactivated Optional expansion of category 'Not effective'	Not in place, turned-off, deactivated. Can also be used to differentiate a local system not having this barrier that differs from corporate standards.	Black

are in a degraded state. The third part of the figure shows 'with barrier management' in place to return risk to the original target value. Barrier 1 is repaired, but since parts are not currently available for barrier 3, a new barrier 4 with equivalent functionality has been installed temporarily. This is a qualitative process; if semi-quantitative support is required then a LOPA assessment might be used. There are management and communication issues associated with adding new barriers and often it might be better to strengthen barriers 1 and 2 to achieve the desired risk level.

5.4.5 The Addition or Removal of Barriers

It is often better to resolve deficiencies in existing barriers (for example, by enhancing its supporting degradation controls, or recognizing the role of previously unidentified degradation factors and implementing effective controls against them) than to add new barriers. That is particularly true if the new barrier will be subject to the same degradation factors that reduced the performance of the existing barriers. This topic is discussed in more detail in Chapter 6.

Are additional barriers required? Additional barriers are normally only required when there has been a change, such as when process conditions change, the vulnerabilities increase, or when new technology becomes available and is deemed to reduce risks in a practical manner or to ALARP. Periodic reviews are necessary to ensure changes are identified as the whole safety management system is a dynamic, constantly changing system. Changes prompting additional barriers can include:

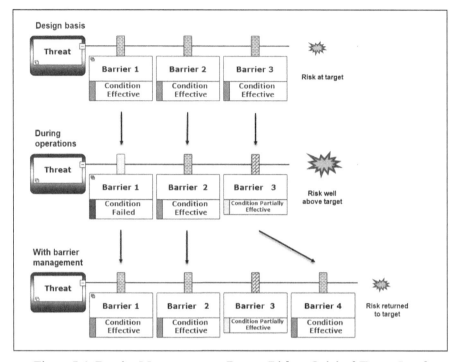

Figure 5-1. Barrier Management to Return Risk to Original Target Level

- an increase in the risk, e.g., changes in process conditions or potential consequence severity as additional people become now exposed to the hazard;
- an existing barrier is degraded and cannot be rectified;
- a review PHA study has identified a deficiency in barriers or degradation controls;
- an incident or near miss has occurred in the plant indicating the existing barriers are inadequate; or
- an incident has occurred at another site that indicates the current design practices are insufficient (e.g., a CSB recommendation of lessons learned for the industry).

Alternatively, an existing barrier could be strengthened, e.g., Safety Instrumented System changed from SIL-1 to SIL-2 or piping material changed to a corrosion resistant alloy from carbon steel. The bow tie can be used for qualitative decisions on these changes, but LOPA or other techniques would support a semi-quantitative decision.

Every extra barrier adds complexity and must be managed through its lifetime (inspection, maintenance, training, etc.). This extra effort might detract from other safety issues on the facility.

Are fewer barriers acceptable? A converse position to additional barriers is whether it might be possible to remove barriers. Examples where fewer barriers might be appropriate include a change to low sulfur crude feedstock no longer causing a hydrogen sulfide risk, or the removal of nearby occupied buildings reducing exposure levels, etc.

As for the previous discussion on additional barriers, if more than a qualitative basis for the decision is required, then a LOPA or similar study would be useful.

Critically, and obviously for a well-managed facility, all changes to barriers need to be assessed as part of an MOC system.

5.5   IDENTIFICATION OF SAFETY CRITICAL INFORMATION

Bow tie diagrams help to capture and identify safety critical information as well as to show important inter-relationships. A key use of criticality information from bow tie analysis is to prioritize the integrity program. The most important barriers can be given greater attention for inspection and testing together with higher priority for repair. The topic of safety critical information, however, is broader than only bow tie barriers and needs careful attention in its own right.

5.5.1 Critical Barriers

In the following text, the term 'critical' includes protection of people, assets, the environment, and reputation, the four commonly used consequence categories in MAE bow ties. Some companies prefer the term safety critical barriers. Regulators focus on consequences for people and the environment.

Barrier Criticality. A barrier may be deemed 'critical' for several reasons (of which more than one may apply in any situation):

a) The threat to which the barrier contributes has a high probability of causing the top event;

b) The consequence to which the barrier contributes is very severe or a risk matrix considering all of the threats has ranked it as high risk;

c) The pathway is of medium risk but has very few barriers (i.e., weak defense in depth);

d) The barrier appears on multiple pathways or bow ties and thus is of cumulative importance;

e) The other barriers on the pathway are known to have a common mode of failure (in principle all barriers on a pathway should be independent, but in practice this is not always possible);

f) It is on a published list of safety critical systems issued as an industry code of practice (e.g., various API Recommended Practice 14C for offshore, API Standard 521 for relief systems); and

g) Expert judgment.

The basis for determining barrier criticality should be considered for each organization, and whether the above set or another is adopted, consistency is important. The agreed basis for determining barrier criticality should therefore be made available for use in all risk workshops and studies.

Barriers that are highly critical often receive high priority for the investment of resources. It is important that not every barrier is labeled as being critical, as then prioritization of operations and maintenance resources cannot be differentiated or achieved.

Once a barrier is determined to be critical, then the degradation controls supporting that barrier should be defined and displayed along the degradation pathway leading to that barrier. This raises awareness of the degradation factors and the importance of the degradation controls to achieving the desired effectiveness of the critical barrier.

Once the bow tie diagram has been verified as correct (i.e., barriers all meet the formal validity criteria and degradation controls are correctly located on degradation pathways), then all the barriers shown are important as their failure makes a major accident more likely. For that reason, many practitioners dislike the term 'critical barrier' or 'safety critical barrier' as it implies other barriers are not critical. However, some barriers can be more important than others and frequently some barrier components are designated SCEs and their associated human actions as safety critical tasks. As noted before safety includes environmental aspects, but some jurisdictions prefer the use of the designation SECE – safety and environmental critical elements.

The regulations in many countries require safety critical elements to be identified:

- Initially in the UK, and now in Australia and all EU countries, safety critical elements (or similarly worded items) must be defined and have performance standards and a system to keep them at this standard through the lifecycle of the barrier. Points made in these regulations include the following: a) 'Critical to the control of major hazards is the correct identification of the major hazard risk control measures (SCEs) and the performance required of them' and b) 'Risk control is achieved through the maintenance and inspection of SCEs to ensure their correct operation, management of change and the management of occasions when the SCEs are impaired'.

- US requirements are less formal, but the CSB and OSHA have expressed interest in identification of safety critical equipment.

The purpose of developing such a list of SCE is that the equipment involved would be ranked higher on the integrity priority program for inspection, maintenance and repair, together with measures such as holding parts in stock to reduce the time for repair. In addition, regulators and management would focus on the status of this equipment during visits.

SCEs are designated to receive prioritized attention. There is however a practical resource constraint: if too many equipment items are designated 'critical', then it is not possible to direct special treatment to them all.

A large facility has many competing priorities and designating a high percentage of items SCE might negatively impact other priorities by stripping away resources and management attention. There is published guidance on typical offshore safety and environmental barriers / SCE (e.g., NORSOK (2008), EI (Energy Institute, 2007)) and more general advice in the CCPS LOPA Guideline (CCPS, 2001). Examples include most safety systems (such as area flammable monitors), fire protection systems (such as deluge or underground systems) and integrity of equipment handling high pressures or large volumes.

In bow tie terms, some barriers, particularly on the consequence side, contribute more to interrupting a bow tie pathway leading to a major accident event than others. Water curtains that can knock down ammonia releases are clearly critical but their effectiveness is not 100% as they only work when the wind direction is towards the curtain and even then, they are not 100% effective in absorbing all the toxic gas. Additional barriers are likely to be required to manage the risk.

5.5.2 Critical Tasks

Critical tasks are activities essential to activate or maintain the barriers that are intended to be in place against the MAEs. These are observations, decisions and tasks performed by personnel during operations or maintenance, which play a direct role in prevention or mitigation barrier functions. Additional activities / tasks can be designated critical because failure to perform the activity correctly can cause or contribute substantially to the initiation of an accident; for example, a valve being left in the incorrect position or activities in an operating procedure executed in the wrong order. It is important that critical tasks are identified, and that people know what they need to do and the potential consequences if the task is not correctly executed (see also CCPS, 2011).

In the context of barrier management, critical tasks can also be directly related to a degradation control against a degradation factor. Maintenance tasks can also be critical; for example, a degradation factor might be maintenance personnel leaving incorrect positioning of isolation valves around pressure relief valves after maintenance. The degradation control might be for an independent operations person to check the valve position prior to accepting the system back in service. Other maintenance tasks such as removing corrosion products from sprinkler heads would not be considered critical if there was a functional system test of the sprinklers and the system is designed allowing for a percentage of heads not to be operational. Operational or maintenance activities or tasks are designated critical if the barrier function is prevented through incorrect or lack of execution of the task.

Mature organizations will have a long list of existing operating procedures. These will often have been developed because of the organization's previous identification of the importance of specific steps, i.e., of critical activities. Critical tasks associated with barriers will often not need to be developed, but rather to be assigned to the existing operating procedures to the barriers.

Organizations may also define 'critical procedures' which describe how to execute a critical task. Critical procedures are attached to a critical task. Organizations may also specify the competency required to conduct the task.

Similar to critical barriers, critical tasks are designated to receive prioritized attention. There is, however, a balance in that if too many tasks are designated critical then it may not be possible to direct special treatment to them all. Critical activities / tasks are part of barrier management (see Chapter 6).

5.5.3 Acceptance Criteria

Acceptance criteria are measurable standards that are used to establish whether a barrier is functioning properly or is degraded to some degree, or in the design phase meets the original design specification. Acceptance criteria are also called performance standards. Acceptance standards during the operations phase can be simpler than full confirmation of the detailed design specification.

Acceptance criteria are the required important performance characteristics of a barrier. For active hardware barriers, this would normally be the output – for example, a firewater flow rate delivered within a set period, or that an ESD closes completely in a nominated time and is below a specified leak rate. For passive barriers, it may or may not be the direct function. For example, a process fluid containment system is easily measured in terms of pipe or vessel wall thickness using common inspection equipment. However, the actual resistance of a 60-minute firewall can be measured on a sample during manufacture, but cannot be measured during operations. Instead, the acceptance criteria might be 'no visible cracks or delamination larger than (some specified dimension)'.

The performance of human and organizational barriers can be hard to measure, although human barriers can be effectively assessed and evaluated.

There are several KPIs that would indicate whether the barrier is working satisfactorily. For example, the work permit and Lock-out Tag-out degradation controls are well documented procedural systems and the paperwork would normally be verified periodically to ensure that the documents have been properly filled-in, that safety steps or checks are properly signed-off, and that close-out has occurred. Other procedures without checklists or sign-offs would require effective supervision or task observation and results of these can be assessed. Another useful form of assessment of human performance would be emergency drills with criteria set and measured for time to muster. The treatment of human barriers is discussed more fully in Chapter 4.

KPIs are discussed in API RP 754 (API, 2016), IOGP 456 (IOGP, 2011) and ICCA Guidance (ICCA,2016), for example. KPIs can track leading or lagging metrics. While many are lagging, there are some that are good leading indicators (e.g., Tier 3 and Tier 4 statistics in API 754). Tier 3 is a demand on, or failure of, a safety system or barrier (e.g., demands on a safety instrumented system). Tier 4 relates to the safety management system supporting safety systems or degradation controls (e.g., percentage of PSV tests conducted on time). The IChemE (2015) and IOGP 556 (IOGP, 2016) have provided additional details on leading metrics. Many companies have developed detailed KPIs for these tiers.

As part of the briefing to the workforce on barriers, especially the ones they operate or are responsible for, they must understand what the acceptance criteria are so they will be in a position to detect if a potential degradation has occurred. This requires that the criteria be clear and directly assessed. The staff may not do the assessment, but they should know how it is done: for example, that relief valves are removed and bench tested according to a specified frequency, or that scaffolds are tested and tagged before use.

Acceptance criteria can be differentiated from formal regulatory specified 'performance standards' which also specify standards for barriers. For example, (referring to Table 5-5) survivability against blast or fire load is one performance standard. The required design strength of equipment of a blast / firewall is derived from MAE loads at design time, but is not measured during operations.

A full discussion of performance standards is beyond the scope of this concept book, as it requires addressing specific regulatory requirements. They are specified in the 1995 UK PFEER Regulation (Prevention of Fire and Explosion, and Emergency Response on Offshore Installations) with associated HSE Guidance documents, by the EI (Energy Institute, 2007 for safety critical and 2012 for environmental critical), and by the Australian offshore regulator (NOPSEMA, 2012).

A short summary of the information typically required is provided in Table 5-5. These normally only apply to barriers, not degradation controls.

Table 5-5. Performance Standard Parameters

Performance Standard Parameters	Meaning	Example
Functionality (Operation)	What is the safety barrier's purpose? Specifies what it must do or should achieve, normally the output.	The fire water system shall deliver, for example, 5000 gpm at 120 psi at the most remote location, within 3 minutes of actuation.
Functionality (Design confirmation)	How is the safety barrier designed / robust against degradation?	The fire water system shall have two independent water supplies with two fire water pumps and with diverse power supplies delivered through a ring main, which can isolate any leak.
Availability	How often will the safety barrier be required to operate satisfactorily? Specifies whether it will be ready to perform and function.	The fire water system requires it to be taken out of service weekly for 2 hours testing; this is 104 hours / year when it is 'not' available.
Reliability	How likely will the safety barrier perform on demand? Specifies whether it will perform the required function on demand and that it will run for a specified duration (i.e., not just start).	The fire pump should start 95% of the time (19 out of 20 actuations). Care is required in calculating the overall system reliability, as variables could include human intervention in the detect – decide – act sequence, or the system may involve other fallible components.
Survivability (Design criteria)	How long will the system perform post-event from effects of MAEs (fire, explosion, vibration etc.)? Specifies for how long the safety barrier will continue to perform the required function on demand. This is primarily for mitigation barriers.	A fire water deluge system should be designed to withstand maximum specific blast overpressure to which it may be exposed. Checks during operation, for example, would be the condition of any blast resistant attachments for deluge sets.

5.6 CONCLUSIONS

This chapter has described some of the basic uses of bow ties. The diagrams are suitable for a broad range of risk management tasks and to communicate to a wide variety of audiences. The primary use of bow ties is to provide a systematic qualitative structure for understanding the major accident event risks associated with a facility and how barriers and degradation controls are deployed to prevent or mitigate these risks. It provides a visual communication tool for operational, maintenance and management staff, contractors, board level management, regulators, and the public. It does this in a manner that is more easily understood than most other risk assessment tools. Bow tie software can display more or less information depending on the needs of the audience. For example, degradation factors or metadata can be hidden for basic communication purposes, but fully displayed to barrier owners or process safety specialists.

Bow ties can make a significant contribution to many other risk management activities in both design and operations. The focus of bow ties on MAE events and on operations with an emphasis on potential barrier degradations differentiates them from other risk tools such as LOPA and QRA which have a greater, though not exclusive, focus on design.

While all barriers in bow tie diagrams are important, some may be more important than others, and so critical barriers should be identified and highlighted. These would normally have degradation factors developed documenting the degradation controls deployed to enhance the strength of the barriers. Guidance is provided on how critical barriers might be selected. However, care is required not to make this list too long as it is not possible to provide additional attention if the list includes almost everything. Guidance for critical tasks follows a similar strategy.

Bow tie diagrams provide an effective basis for the auditing, inspection and maintenance of barriers.

6

BARRIER MANAGEMENT PROGRAM

6.1 BARRIER MANAGEMENT STRATEGY

Effective risk control requires there to be a continuous barrier management process to monitor their operation and to verify that barriers are not degraded. Technical barriers are normally managed through the asset integrity element of the facility process safety management program. Human barriers are often not rigorously managed. Bow ties help to make the barrier management process more transparent and ensure that all barriers are well managed.

Key topics covered in this chapter include:

- The need for a barrier management strategy;
- The link to process safety management programs;
- How to develop a barrier management program;
- Direct monitoring of hardware and human barriers;
- Indirect indicators of barrier performance;
- Management of change;
- Barrier life cycle management; and
- Updating bow tie diagrams.

6.1.1 The Need for a Barrier Management Strategy

There is a need for top level management understanding of major accident risks and how these are managed safely with specific barriers. This helps senior managers understand the importance of ongoing support activities to manage the barriers and their degradation controls effectively.

As soon as a facility is commissioned, the effectiveness of its barriers inevitably begins to deteriorate from the initial or intended design effectiveness due to barrier degradation. There are many examples of process facilities that were designed with effective barriers in place, but where those barriers had been allowed to progressively degrade over time. Several of these are discussed in the CCPS book 'Incidents that Define Process Safety' (CCPS, 2008c). A list of major incidents highlighting important barriers and / or degradation control failures is shown in Table 6-1. (Note that 'Incidents that Define Process Safety' included more examples of barrier failures than shown on Table 6-1, but space limits the examples to the more important ones). These are not isolated examples; almost all major incidents can be traced to failed barriers and the table could have been much longer.

A useful guideline on holistic barrier management is available from the Norwegian Offshore Regulator (PSA, 2013 – freely available on their website). This non-mandatory guide outlines a barrier management strategy for a Norwegian offshore oil and gas facility, though many of its ideas are applicable to barrier management in any application.

All the events in Table 6-1 could have been prevented or significantly mitigated if the barriers or degradation controls which had been installed or were part of normal operations had been working properly at the time of the incident. It is for these reasons that ongoing barrier management is vital.

The Texas City accident highlights an important issue for managing barriers, and this is changing norms. When the local atmospheric vent was installed, it was consistent with common practice, though industry practices had changed over the years, and venting to a flare had become the established good practice. In that incident, the issue for barrier management was not whether the atmospheric vent barrier was working, but whether the barrier itself met current practices, i.e., was it time to upgrade to current practice to collect local reliefs and direct these to a flare? (This issue is discussed more fully in Section 6.2.6 on life cycle issues).

Table 6-1. Major Incidents with Barrier or Degradation Control Failures that should have Prevented or Mitigated the Event

Event	Top Event and / or Consequence	Example of important Barrier or Degradation Control Failures
Bhopal, India 1984	Toxic methyl isocyanate (MIC) release, fatalities	MIC refrigeration system permanently shut down allowing runaway reaction to occur. Vent absorber and flare not operating.
Piper Alpha, UK 1988	Gas condensate release, fire, escalation, fatalities	Work permit system, automatic deluge on manual which inhibited initial fire response, and temporary refuge were all degraded.
P-36 offshore rig, Brazil 2001	Two (2) explosions in leg, sinking, fatalities	Improper maintenance isolation of drain tank allowed ingress of pressurized hydrocarbons, causing a pressure burst followed by a hydrocarbon explosion.
Texas City refinery, USA 2005	Vent release of gasoline, vapor cloud explosion, fatalities	The atmospheric vent location in plant no longer met industry norms to discharge to a flare (the vent filled with liquid that spilled into the plant), separation of people

Table 6-1. Major Incidents with Barrier or Degradation Control Failures that should have Prevented or Mitigated the Event, continued

Event	Top Event and / or consequence	Example of important Barrier or Degradation cControl Failures
		(temporary buildings were sited too close to hazard). There were as well deeper maintenance backlog and safety culture issues (examples of multi-level degradation controls).
Buncefield, UK 2005	Tank overfill, spill, vapor cloud explosion, major fire, groundwater contamination, injuries	Tank level gauge barrier failed (it was known to suffer intermittent failures). The independent high level switch / trip barrier was left in the wrong operating mode when it was overhauled three months before the event. Dikes failed to contain burning gasoline when dike wall joints melted after fire water applied.
Georgia sugar refinery, USA 2008	Dust explosion in sugar plant, fatalities	The barrier of allowable sugar dust thickness (as recommended by NFPA) was exceeded by many times permitting a dust cloud to form that exploded.
Deepwater Horizon, USA 2010	Blowout event, lengthy oil spill event, fatalities	Deepwater well control barriers, the BOP, and ignition control all failed, and the flow diverter was not used correctly.
LaPorte TX, USA 2014	Toxic gas release inside building, fatalities	Methyl mercaptan release – ventilation system barrier not to design specification; fans not working on the day.

6.1.2 Linkage with the Process Safety Management Program

Bow ties assist with the understanding of major accident risks and demonstrate the role barriers and degradation controls play in managing risk. The bow tie diagram shows how barriers can degrade and how the support of barriers links to the overall facility management system (Figure 6-1). Degradation controls and barriers can be mapped to parts in the management system which serve to ensure barriers are effective such as operating procedures, training, competence, maintenance, inspection, and auditing.

Most barriers rely on many components of the management system (e.g., training and competence affects every barrier, or at least their degradation controls) and linking the key parts of the management system associated with each barrier can highlight gaps in the management system. The Australian regulator (NOPSEMA) highlights:

'...the safety management system needs to be shown to fully support and maintain the performance standards of the control measures within an integrated management framework'.

The mapping described here helps demonstrate this. But to avoid duplication of effort, care is required to ensure the bow tie process does not replicate the process safety management system already in place for audits and checks.

Many aspects of the bow tie diagram can be linked to specific management system elements. For example, a pressure relief barrier and its degradation controls can be linked to training and maintenance and inspection programs for its direct support, and auditing for checking that these have happened correctly. Typically, there will be several links for each barrier and its associated degradation controls. The linkage can identify gaps. For example, in a hardware + human barrier such as 'gas detection, human decision, and ESD system', there needs to be some element in the training program that addresses this critical human decision and action to press the ESD button. This can also link to the Stop Work Authority degradation control, which might be displayed on a multi-level bow tie.

6.2   BARRIER AND DEGRADATION CONTROL MANAGEMENT PROGRAM

6.2.1 Definition of the Program

Barrier and degradation control management should form part of the overall process safety management system. Barrier and degradation control management can be integrated into the facility integrity system and HOF programs. The list of barriers and degradation controls should be collected from the bow tie diagrams, as well as relevant metadata (e.g., if the team has designated any of these as contributing more to prevention or mitigation than others). Hardware items would fall into the asset integrity system, whereas human related barriers or degradation controls require links to specific procedures, training and more general human performance programs.

A typical flowchart for a barrier management program is shown in Figure 6-1. This has similarities to steps in the CCPS book 'Guidelines for Mechanical Integrity Systems' (CCPS, 2006), but with some extra steps added for clarity and to more directly match barrier performance.

The Barrier Management Program is itself a management system element. Like all system elements it requires a policy, plan, implementation, measurement, audit and management review. Bow ties are used within the safety management system to support the management of barriers and degradation controls. The flowchart in Figure 6-1 shows some key feedback loops where the bow tie structure or barrier monitoring intervals may need updating. The same process can be applied to degradation controls.

Ongoing process safety requires good leadership and a suitable assurance program addressing all elements of the process safety management program. The barrier management program should become an element of the overall process safety management program to monitor deterioration of specific barriers and to direct attention to returning these to their desired standard.

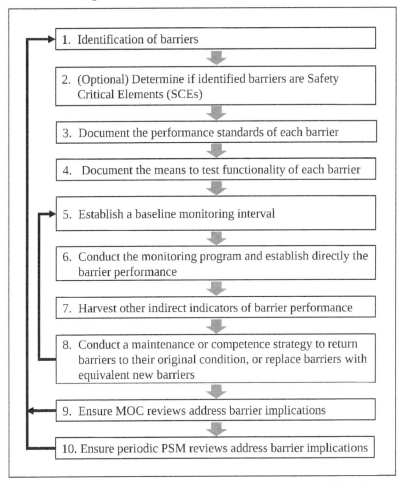

Figure 6-1. Barrier or Degradation Control Management Flowchart

Step 1 'Identification of barriers' identifies the barriers to be covered by the barrier management program. This list comes directly from the bow tie diagrams created for the facility.

Step 2 'Determine if barrier is a safety critical element' is optional or may be required by some companies or regulatory agencies. The basis for determination of criticality was discussed in Section 5.5 which identified several criteria that may be used. The advantage of carrying out Step 2 is that it allows a more flexible and cost effective maintenance strategy to be developed as not all barriers have the same importance; although, as was emphasized in Chapter 5, in principle all barriers are important, not just a subset, and need to be managed.

Step 3 'Document the performance standards for each barrier'. These performance standards are the levels that deliver the agreed risk target level. It is not possible to assess barrier performance if the performance standards are not defined. Generally, performance standards are defined for hardware items. Procedure-based degradation controls are known to have faults, but it is less common to nominate a target fault rate for work permits or LOTO (Lock-out Tag-out). In principle, there is no reason that human based barriers should not also have performance standards as do hardware barriers, and this is likely to be an area of improvement in the future. The Chartered Institute for Ergonomics and Human Factors has provided guidance and examples of how a human performance standard can be specified (CIEHF, 2016).

Step 4 'Document the means to test the functionality of each barrier'. For some barriers, this is straightforward and may be done while the plant is running (e.g., testing the functionality of gas detectors), whereas others are more difficult or may require a shutdown (relief valves must be removed and taken to a workshop). In some cases, it will be possible to use a partial test. For example, ESD valves can be tested via a partial stroke to 10 - 15% closure to see whether the valve moves on command, but this does not test the actuator time to fully close or the valve leak integrity. Testing human barriers can be more difficult but direct task observation and audits of documentation, among other approaches, can be used. This topic was discussed in more detail in Chapter 4 (Section 4.4).

Step 5 'Establish the baseline monitoring interval'. Generally, the interval would follow the manufacturer's recommendations for the item, but tests that are more frequent may be required, such as where a LOPA study has nominated a higher SIL level or level of reliability. Where a safety case is used (either voluntary or because of regulation), a written maintenance and inspection scheme may need to be developed that defines the tests and intervals.

Step 6 'Conduct the barrier monitoring program' is the implementation of the program and collection of barrier performance data. This would be supplemented with other direct sources of information, such as from incident or near miss investigations (see Chapter 7 on advanced uses of bow ties), from audits, from supervisor observations or management tours. Operational success as well as failure are both relevant. The Norwegian PSA documents successes of barriers as well as failures in their incident investigations. This also matches the idea of focusing more on barrier success pathways as discussed by EuroControl in their white paper: 'From Safety-I to Safety-II' (Hollnagel & Leonhardt, 2013). Another means of testing the performance of barriers is to log barrier failures and successes for every loss of primary containment in a facility corresponding to Tier 1, Tier 2 or Tier 3 events (as per API 754). Every significant release means that the prevention barriers failed and a top event has occurred (i.e., a loss of containment event), but at least one of the mitigation barriers worked thereby preventing an MAE.

Step 7 'Harvest other indirect indicators of barrier performance'. This might include inspection or maintenance backlogs on specific barriers. It may also be human factors audits or inspections that warn of wider human barrier degradation, etc. The data from Steps 6 and 7 can be combined. The CCPS addressed this topic in a paper 'Process Safety Leading Indicators Survey' (CCPS, 2013b). It is important to track indicators for human barriers and this links well to current human factors ideas (e.g., Hudson & Hudson, 2015, CIEHF 2016). Computerized maintenance management systems can be an aid to implementing Steps 6 and 7 as these track a wide variety of relevant information.

Step 8 'Return barriers to their original condition' is the implementation of activities to return barriers to their design performance. It might involve a direct repair for hardware, a review of procedural degradation controls, or implementation of a replacement barrier if it is difficult to repair the original. If Step 2 has been carried out, then a risk-based priority plan can be developed that optimizes the response in terms of both safety and facility economics, similar to the achievement of Risk Based Inspection (API 580 / 581) which has delivered better safety at lower cost to refining and offshore facilities around the world.

Step 9 'Ensure MOC reviews address barrier implications'. This step addresses MOC activities and ensures that these address any implications of change on barrier effectiveness. This topic is addressed in greater detail in Section 6.2.5.

Step 10 'Ensure periodic PSM reviews address barrier implications' is similar but it addresses longer-term changes that might have been missed by the MOC process or slower changes that show that current barriers no longer meet industry norms (e.g., atmospheric vents vs. enclosed flare systems; manual shutdown vs. automated SIL-rated ESD systems).

6.2.2 Monitoring of Hardware Barriers

Barrier types were categorized into five convenient groups (with definitions shown in Table 2-9):

1. Passive hardware, e.g., fire wall, tank dike, drainage system;

2. Active hardware, e.g., relief valve, firewater deluge;

3. Active hardware + human, e.g., ESD system actuated by operator in response to gas detector alarms;

4. Active human, e.g., operator continuous observation of a process; and

5. Continuous hardware, e.g., ventilation system, cathodic protection.

Hardware barriers can be subdivided into subsets of safety critical elements, either for increased focus or due to regulatory requirements. The monitoring interval for these is variable: a passive blast wall has a very slow degradation period, whereas a gas detection system may degrade quickly. Hence, these have different monitoring intervals, but both may be critical.

Monitoring hardware barriers is generally straightforward in concept, although in practice, it can be difficult or expensive, and many barriers can only be measured during shutdowns. Typical techniques (which notably all rely on human performance) include:

*   Visual observation, e.g., emergency lighting, obstructed escape ways, fire extinguishers beyond service dates, passive barrier condition, inhibitors that remain in place beyond requirements, locked-open valves before reliefs;

*   Inspection during operations, e.g., gas detector accuracy, fire water deluge rates, ultrasonic pipe wall thickness measurements, ventilation flowrates;

*   Inspection during shutdowns, e.g., tank bottom thickness measurements, relief valves; and

*   Preventive maintenance programs, e.g., lubrication, calibration, periodic seal replacement, and other tasks performed to prevent faults during operation.

In bow tie terms, visual observation, inspection and maintenance can appear as degradation controls on degradation factor pathways as they prevent impairment of the main pathway barrier.

All this data should be entered into a monitoring system, generally a computer-based maintenance management system (CMMS). This allows for tracking degraded conditions before a barrier fails its performance standard. A classic example is the containment barrier of metal wall thickness. Inspection records on thickness will show a trend that can allow confidence in the barrier to be determined and for monitoring intervals to be adjusted (i.e., to sooner or later than originally planned, as per Steps 6, 7, and 8 and back to Step 5 in Figure 6-1). This shows the close linkage of the asset integrity system to preventing the degradation of barriers. It is not just failures that are of interest; but also the trend in barrier condition to display this type of steady degradation.

6.2.3 Monitoring of Active Human Barriers

These types of barriers involve human activity. While this is often a strength (humans are more flexible in response to unusual circumstances and often at lower cost than equivalent hardware), it makes these barriers subject to human failure modes. This topic was discussed more fully in Chapter 4.

Generally, there will be management system components to ensure a workforce that is competent and fit for duty; for example, assessing job demands, stress counseling, medical assessment, fatigue management, drug and alcohol testing, and safety culture monitoring. These programs are controlled by the management system, although it is important that the organization recognizes the role these degradation controls play in the robustness of the overall barrier management system. In bow ties, some of these management system elements can be shown as degradation controls: training, competence, task observation and supervision for example. Other deeper level degradation controls (i.e., degradation controls supporting degradation controls) would be displayed as extensions in multi-level bow ties (see Chapter 4). An example might be the degradation control of 'task monitoring', itself supported by a control 'positive safety culture': a culture that understands the importance of task monitoring and encourages intervention if signs of possible problems are detected or there are concerns over safety.

There are many approaches to assessing the effectiveness of human barriers. While some are relatively straightforward (such as task observation, documentation, audits, and incident reviews), others can be more complex, sometimes requiring measurement. Observations can provide confidence that critical tasks are correctly executed, even long after training is completed. Work permits and Lock-out Tag-out have forms that can be checked for correct completion. If persistent errors are found, then a safety publicity campaign, re-design of the forms or remedial training could be arranged. The same applies for critical start-ups: completed checklists enhance the activity and allow for post-activity reviews. Incident and near miss reports can also be useful to establish if the human barriers worked effectively or if there were failures.

An indirect way to measure human-related barriers is by using KPIs for human performance such as process safety indicators from API 754 (API, 2016). Tiers 3 and 4 relate to demands on safety barriers and safety management system deficiencies. Many of these directly relate to human performance. CIEHF (2016) has discussed the use of leading indicators for human factors aspects of barrier systems and illustrates how measurable performance criteria for human barriers can be defined.

An issue with 'active human' and 'active hardware + human' barriers is that in major accident events, when there is high stress and danger, individuals may make errors or omit steps in critical tasks. The CSB report of the Deepwater Horizon accident gives examples where key active human barriers were not actuated correctly or in sufficient time (CSB, 2016). Conversely, successful safety actions terminating an accident sequence due to greater flexibility and overall competence of human barriers are not well reported. This highlights the ideas discussed in Section 4.1.3.

Hardware barriers tend to be more predictable in specific duties, but whether they are more reliable than humans, given human ability to react to unexpected circumstances, is open to debate. It is generally recommended that higher risk pathways have some hardware barriers, and do not to rely solely on human actions. However, where higher risk pathways only have human barriers, then measures to support these through critical task analysis and monitoring, human factors analysis and bow ties assume greater importance.

6.2.4 Audits, Incidents, Regulatory Findings, Culture Programs

Auditing of management system barrier activities and barrier performance data provides assurance of the measures in place. Bow tie barrier reviews should be conducted regularly by line management; audits are a separate activity and generally conducted by people outside of line management.

One approach to line management reviews would be to perform a review of one of the bow ties on a unit during regular (monthly) management meetings. This would serve to both check the status of the barriers and degradation controls originally proposed but also to onboard new members of the local management team as to the barriers and degradation controls that are relied on to prevent MAEs.

Audits normally apply to the safety management system, but since this supports barrier operation and includes the asset integrity element, audit findings do give an indication of barrier reliability. For example, if the audit finds multiple examples of overdue inspections, this may be an indicator of undetected hardware problems. Notes can be added to the bow tie diagram so that it represents the most current situation and removed when the issue is resolved. This is more practical when using bow tie software.

Incidents are a valuable direct measure of barrier performance for those barriers involved. Incidents or near-misses cannot occur unless some barriers have failed. Collecting this information from incident records and near miss reports, or using a barrier based investigation technique can be very powerful (see Chapter 7). Often inspections or training records may indicate all is working well for barriers, but in an incident, the barrier has failed. This can be due to random failures, but it may indicate that formal measurements are not accurate. Part of the problem is that human error is situational: it nearly always occurs in the context of the specific situation, preceding events, and expectations at the time. Inspections and training records rarely capture or even consider the situational nature of human performance. Software tools allow the addition of notes concerning barrier failures from incidents to be shown on a bow tie diagram alongside all occurrences of that barrier. This failure history can be powerful information for readers of the diagram.

Regulatory findings, which may occur after a regulator visit, can also provide useful information. Regulators see many different facilities, and they often visit with specific topics to address which have caused problems elsewhere. Their findings can be harvested for barrier degradation information.

Safety observation programs, such as the Hearts and Minds Toolkit (Energy Institute, 2016) can provide an indication of staff safety culture. If this is combined with observations on Conduct of Operations and Operational Discipline (CCPS, 2011), then a broader picture can be built of the extent to which the generic human and organizational safeguards intended to be in place have actually been implemented. The CIEHF white paper on 'Human Factors in Barrier Management' includes a set of questions that can be used to assess the organizational culture surrounding barrier performance (CIEHF, 2016).

6.2.5 Management of Change

Changes occur during design, construction, operations and maintenance of process facilities. Unintended or unanticipated consequences of change have frequently been identified as a key issue in many incidents (e.g., Flixborough 1976 in CCPS, 2008c). Management of Change (MOC) is an element of the facility process safety management program. A CCPS Guideline book directly addresses this topic (CCPS, 2008b). Organizational change also affects safety and a Management of Organization Change (MOOC) process is recommended to address those kinds of changes (Guidelines for Managing Process Safety Risks during Organizational Change, CCPS, 2013a). The EI addresses these topics in its process safety management framework documentation on its website.

It can be especially challenging for a MOC program to identify potential impacts on how well human barriers will perform in the future. There have been many cases where, because changes have made work more difficult, or made it more difficult to comply with procedures, people have eventually found short-cuts or easier ways of working that have defeated barriers: this tendency to find an easier way of doing things is deeply embedded in human nature. An MOC program should specifically ask whether the proposed change would make any critical tasks more difficult or make it more difficult to comply with any steps in a critical procedure.

The MOC process is usually well defined and involves some form of risk assessment or hazard analysis and several review steps. Since changes can affect barrier operation, the MOC process may itself need some modification to ensure impacts on barrier effectiveness are adequately considered in the risk assessment so that any weaknesses introduced are compensated for. The Bhopal incident (Table 6-1) is an example where actions were taken to disable key prevention and mitigation barriers to save costs, but without an assessment of the safety impact; these should have been subject to an MOC review.

Organizational changes can be more difficult to recognize than physical changes to barriers but can still have a large effect. A well-known incident that emphasizes the unanticipated effects of organizational change is the gas explosion that occurred at Longford in Australia in 1990 (Hopkins, 2000). An onshore gas processing facility suffered an LPG breakthrough due to long-term well fluid composition change when a heat exchanger became cooled to a temperature at which icing and embrittlement occurred. The process engineers, who might have been able to diagnose the issue and warn of metal embrittlement, had all been moved to the head office around 130 miles away. The remaining staff, although well experienced, did not have the same degree of metallurgical and process safety knowledge. No longer having easy access to the process engineers, they made the decision themselves to introduce hot oil to solve the icing problem. Unfortunately, this caused mechanical stress, and the embrittled steel failed, releasing the contents. Two staff members were killed, eight were injured, and a long-lasting fire broke out at a critical facility location. The city of Melbourne and some regional areas in Victoria lost their gas supply for nineteen days with major economic impact. (Some additional details of this incident relevant to life cycle issues are discussed in the next section.)

6.2.6 Barrier Life Cycle Management

All facilities go through a life cycle and it is important that barriers, like all other parts of the facility, are managed through these different phases. This life cycle process of identifying and specifying barriers (during the design phase), to installing barriers (during commissioning), and managing and assuring them (during operations) has reached different levels of maturity in different locations depending on regulations, industry or individual organization's policies.

The Longford gas explosion, mentioned in Section 6.2.5, also illustrates how, over a facility's life cycle, 'creeping' changes can affect barriers. In common with many wells, the fluid composition changes over the well lifetime. At Longford, over time the well was gradually producing more LPGs. A separation column in the front-end of the plant was designed to prevent LPG from reaching the back end of the plant, as the materials at the back end were not designed for possible low temperatures associated with LPG flashing. The separation column was also acting as a barrier. Life cycle management needs to address such long-term, 'creeping' changes and determine whether the process and safety barriers can cope with those changes. It is important to recognize that process items can also act as barriers and that the consequences of long-term changes on barrier performance need to be addressed. This is difficult but essential.

One approach to lifecycle management would be to integrate bow tie reviews into the facility's periodic PHA reviews as required by various process safety regulations for high hazard facilities.

6.2.7 Updating Bow Tie Diagrams

Bow tie diagrams also need to be managed through the life cycle of a facility to remain current. Many issues can lead to barriers needing to be modified during the life cycle: replacement parts unavailability, MOC reviews, five-year process safety management updates, or response to incidents. These changes, along with associated metadata (barrier strength, barrier owner, performance specification, etc.), should be entered into the bow tie diagrams. The diagrams are an effective means to communicate such changes to staff at toolbox talks or similar forums.

After the initial implementation of bow tie methodology, companies will likely find gaps or inconsistencies in the first bow ties, usually discovered in process safety reviews. This type of issue occurred after PHA was introduced as a requirement in the US in 1992. Early PHAs had gaps (perhaps giving 80% solutions), and these were much improved during the subsequent periodic reviews (driving towards full identification of hazards). The situation will be the same with bow ties. The first development adds real value, but later revisions will increase this value. One activity to improve the first bow ties would be common facilitator training and peer review. While corporate experts cannot participate in every bow tie session, they can review the bow ties produced and provide suggestions for improvements.

Finally, most bow ties are created in software, and just like PHA or P&ID drafting software, these tools change and are updated over time. Unless files are imported and updated in new software versions, the files may be unreadable when a need arises perhaps ten years later. Periodic opening of older bow ties in newer software versions to confirm readability is an ongoing task. Software provides a wide range of functionality, such as the ability to easily track barrier status or incident investigation features. These encourage using the software regularly and keeping the diagrams up-to-date.

6.3   ORGANIZATIONAL LEARNING

An important benefit of bow ties is that they can enhance organizational learning and sharing of knowledge. This might effectively be achieved with corporate bow tie templates developed centrally and then issued to facilities to adapt to their local conditions. While the corporate bow tie templates provide good guidance, they are usually applied flexibly to account for local conditions. For example, while there may be a standard corporate approach to active fire protection, it might be decided that a pump station in a remote location would not have active fire protection, only passive. This decision would be made based on careful analysis of the local safety needs, including possible workforce or public exposure. Thus, the corporate bow tie is used, although in a flexible manner.

Generic corporate bow ties can be a very powerful means to share lessons learned. Barrier failures leading to improvements at one facility can easily be transferred to the corporate bow ties. These insights can then be generalized so they can lead to system improvements across the organization. Typically, only main pathway barriers would be defined, as degradation controls on degradation pathways can vary across facilities depending on the local context and situation. This allows a good balance, preventing the generic, good practice bow tie templates from becoming overly prescriptive while allowing customization at individual facilities to local circumstances. For example, a generic corporate template might specify active fire protection as a main pathway mitigation barrier, but a local facility might select between hydrants, sprinklers, or deluge and display those as degradation controls along with controls for detection and decision supporting the main pathway mitigation barrier. This concept is shown in Figure 6-2.

The use of bow ties for incident investigation is discussed in Section 7.5. That approach shows how barriers perform in actual incidents and how this can be used for organizational learning. The learning is driven because some important barriers appear in multiple pathways (e.g., work permit systems, pressure relief valves) and if these are found to have been degraded in one incident they may also be similarly degraded wherever they appear. Lessons from an incident on one part of a facility or an organization can have much wider impact. It could be the role of the barrier owner or bow tie owner to ensure those lessons are applied everywhere.

6.4   CONCLUSIONS

The creation of bow tie diagrams should be seen as the start of an ongoing risk management process, not the conclusion. Barriers degrade over time, at different rates for different types of barriers. Evaluation of barrier status will vary depending on barrier type, most likely involving a combination of direct measurement and indirect assessment.

There is a direct link between bow ties and the process safety management program. The latter is required to sustain all the barriers and ensure they continuously operate at their desired performance level. This chapter also reviewed some different evaluation approaches and whether these are direct or indirect.

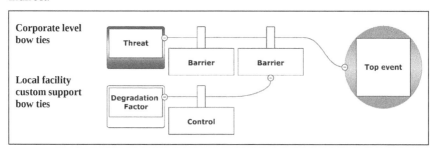

Figure 6-2. Corporate Template and Local Facility Bow Ties

Barrier diagrams need to be updated throughout a facility life cycle to account for changes and improved technology. Bow tie software makes this task easier compared with static paper drawings and modifications can be made easier to identify for users.

7

ADDITIONAL USES OF BOW TIES

7.1 ADDITIONAL USE EXAMPLES

Prior chapters in this concept book have addressed the structural elements of bow ties (Chapter 2), how to create bow ties (Chapter 3), human factors in bow ties (Chapter 4), their basic uses (Chapter 5) and how to manage bow ties and barriers (Chapter 6). The purpose of this chapter is to outline some important additional uses. The additional uses covered in this chapter include:

- Linking bow ties to HAZOP, LOPA and SIL determinations;
- Integrating bow ties into Safety Case and ALARP demonstrations;
- Operationalizing bow ties (e.g., permitted operations based on barrier status (SOOB / MOPO);
- Incident investigation;
- Integration with risk register and other risk tools;
- Bow tie support to real time dashboards;
- Verification programs and barrier health reviews;
- Bow tie chaining; and
- Enterprise wide analysis and window on systemic risks.

7.2 LINKING BOW TIES TO HAZOP, LOPA AND SIL

Other safety studies such as PHA generally, and HAZOP, LOPA and SIL assessments specifically, also document barriers; therefore, it is useful to show how bow tie barrier analysis links to these other studies.

7.2.1 HAZOP

Hazard and Operability Studies (HAZOP) are a key means to identify hazards in a process. The method is described fully by the CCPS (2008) and IChemE HAZOP Guide to Best Practice (Tyler, Crawley, & Preston, 2008). Bow ties are not a hazard identification method. Instead, they further develop specific MAE events identified by the HAZOP. In recording HAZOP, it is standard to document risk controls (covering both barriers and degradation controls in the bow tie context); these are listed to aid the team in determining whether additional control measures might be warranted. This list of controls can be a key input to developing the bow tie diagram. However, there are some caveats. HAZOP addresses the entire process, divided into possibly dozens or hundreds of nodes, and for reasons of time management, the HAZOP team usually cannot spend the same amount of time documenting every barrier on every node as bow tie teams do for a smaller number of specific MAEs. If they did, the duration of the HAZOP study might be increased very significantly. Often teams will stop once they have identified three to five barriers or the set that may be utilized as Independent Protection Layers (IPLs) during the LOPA process. HAZOP teams do not apply the validity criteria for barriers (discussed in Section 2.6.3); degradation controls such as training and asset integrity program frequently appear mixed in with full barriers in the list of HAZOP safeguards. This lack of differentiation is not an error, just a difference in the study approach and scope.

The linkage between HAZOP and bow ties, therefore, is that the HAZOP (or other PHA study) identifies the MAE events and a selection of those is made to develop bow ties to demonstrate the full range of barriers and controls deployed, not just the set identified by the HAZOP team.

7.2.2 LOPA

Layer of Protection Analysis (LOPA) is described fully in a CCPS concept book (CCPS, 2001). LOPA is a simplified, semi-quantitative assessment tool, working generally in order of magnitude approximations. It also starts from the PHA (e.g., HAZOP) and for selected major accident scenarios it applies a quantification procedure to see if the barriers in place reduce risk to the desired target. The specific risk reduction value of a barrier, termed Independent Protection Layer (IPL), was the subject of an additional Guideline (CCPS, 2015). LOPA addresses both hardware and human barriers. It also takes into account conditional modifiers (such as wind direction probability) which are not included in bow tie diagrams (CCPS, 2013c). Some factors such as ignition probability can be both a conditional modifier in LOPA and a barrier (in the form of ignition control barrier) in bow tie diagrams.

Specific requirements in LOPA are that IPLs be effective, independent, and auditable. The CCPS LOPA book provides some examples of barriers that would not normally be considered IPLs (Table 6.1 in CCPS, 2001). These include training and certification, procedures, inspection, maintenance, signs, and fire protection. Many of these do appear in bow tie diagrams but as degradation controls rather than main pathway barriers.

Bow tie diagrams are qualitative, and generally counting of barriers is discouraged as a test for sufficiency of protection. LOPA or full QRA is a better approach for this. LOPA requires an explicit numerical risk acceptance target which has no direct equivalent in bow tie assessments. However, some bow tie software provides full LOPA capability within the bow ties. A comparison of the LOPA and bow tie approaches is provided in Figure 7-1; both examples assess two barriers.

Figure 7-1 shows the difference in approach between LOPA and bow tie for decision making. LOPA is quantitative and repeatable between different teams when the order of magnitude values are extracted from standard sources. However, passing or failing is against a specific target and order of magnitude calculations can sometimes produce results which judgement would not have reached. The bow tie approach is qualitative and relies on team judgement to assess the risk against how similar levels of risk are managed elsewhere on the plant or company. If other models show that two prevention barriers are required, then these are assumed standard strength barriers – some combination of effectiveness and reliability. The team can choose to compensate for a weaker barrier by strengthening the other barrier (as shown in the figure) or adding an additional one. Thus, both approaches allow decisions to be made, but where consistent quantitative results are sought, then this would favor LOPA.

LOPA Risk Assessment	Bow Tie Barrier Assessment
Quantitative assessment of barrier sufficiency	Qualitative assessment of barrier sufficiency

Calculation
Consequence category = C
Risk tolerance criteria R_T (as frequency for C)
Initiating event frequency = F_i
Conditional Modifiers
    Probabilities = P_1, P_2
Independent Protection Layers
    Failure probabilities=IPL_1, IPL_2
Frequency for consequence C, R_{FC} is the product of the above terms
    $R_{FC} = F_i*P_1*P_2*IPL_1*IPL_2$
	Workshop
	Team judges that two barriers are adequate as risk is similar to other pathways which use two barriers.
	During workshop discussion of the two barriers, the second of which is judged of lesser quality (e.g., maintenance in the past shows poor reliability).
	
	Barrier decision
	Judgement is used to enhance the first barrier to be stronger than average to compensate for weakness in the second barrier.

Protection decision Compare R_{FC} to risk target R_T (expressed as frequency for consequence C) If $R_{FC} > R_T$ then greater protection is required: either new or stronger IPL If $R_{FC} < R_T$ the protection meets target and existing IPLs are sufficient R_T values from Table 8.1 of CCPS LOPA Concept Book (CCPS, 2001)	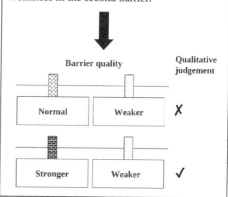

Figure 7-1. Comparison of LOPA and Bow Tie Barrier Assessment

Where qualitative decisions are sufficient, then bow ties provide an easier to communicate decision process.

LOPA and QRA are powerful tools for decision making on the need for potentially costly risk reduction measures with traceable and auditable quantification. However, as may be seen in Figure 7-1, LOPA calculations can become complicated (refer to Appendix B examples in CCPS, 2001) and would be poor for staff communication, whereas the bow tie can be used to present the LOPA barrier quality results in a format which is much easier to explain and understand.

7.2.3 SIL Determination

Safety Integrity Levels (SIL) are defined in standards (such as IEC 61508, 61511 and ISA S84) and these provide a common quantitative definition for the risk reduction potential of a safety function. These are typically 'active hardware' barriers, but may be 'active hardware + human' barriers (e.g., operator initiated ESD) or 'active human', and they appear frequently in LOPA studies as IPLs. The standard defines several levels of SIL – generally SIL-1 to SIL-4 with order of magnitude steps between. SIL-1 is the most common in process facilities with a probability of failure on demand in the range 0.1 – 0.01. SIL-2 reduces these by one order of magnitude and is more difficult but possible to achieve in typical process facilities, typically with extra detectors and voting systems. IEC 61508 does not permit humans to be present in SIL-2 and higher integrity systems so these are active hardware barriers. SIL-3 is complex and often requires three or more voting detectors and multiple actuators (e.g., both close shut-off valve and cut power to a compressor, etc.). Some chemical companies prefer one strong hardware barrier (SIL-3) to several weaker barriers (unrated, SIL-1, or SIL-2), as the fewer number allows for easier management and operational focus. SIL-4 is impractical in process facilities and so the units would be redesigned to avoid their use. The nuclear industry is an exception to this.

Barriers identified in a SIL study for a process facility would normally appear in a bow tie analysis as main pathway barriers. However, bow ties consider a wider range of barriers and degradation controls, including human and organizational controls, than might be permitted in SIL analysis.

7.3   INTEGRATING BOW TIES INTO ALARP DEMONSTRATIONS

Risk assessment and risk management describe the management process for assuring that the risks associated with the facility's operation are maintained at a defined risk level. Risk assessment is an integral part of the CCPS Guidelines for Risk Based Process Safety (CCPS, 2007) and the EI PSM Framework. Good practice operators identify their MAEs and demonstrate that these are adequately controlled to make risk levels meet a defined risk target (in other words, having demonstrated that the risks have been reduced to ALARP). Working to ALARP means that risks are reduced to a level that is 'As Low As Reasonably Practicable' taking into account the magnitude of the consequences and the frequency of the consequence occurring. This is demonstrated by comparing the baseline design risk against the risk after feasible additional barriers, and these measures should be implemented unless the additional costs (in time, trouble, or money) are grossly disproportionate to the risk reduction. ALARP was defined by the UK HSE initially (HSE R2P2, 2001) and is well covered by HSE guidance (HSE, 2013b) and in CCPS (2009). It has been extended to cover environmental aspects as well by the UK downstream sector of industry working with regulators (SEPA (2016a, 2016b). While ALARP is a regulatory requirement in many countries (Britain, Australia, New Zealand, Singapore, etc.), it is a good practice approach adopted by many companies for their global operations regardless of local regulatory requirements. The CCPS (2009) guideline discusses how this is applied, including the idea of gross disproportion.

In bow tie terms:

- ALARP assessment using bow ties is qualitative and would consider the number of barriers, the type and diversity of barriers (hardware, human), their quality / effectiveness, the distribution of barriers between prevention and mitigation functions, and how these match the threat likelihood or the potential consequence magnitude.

- The ALARP test should be applied based on the practicality of adding an additional barrier and the quality of degradation controls against degradation of the barriers. It should not be based on an arbitrary barrier count. Adding barriers also adds complexity and time, which may detract from improving safety elsewhere.

- A good approach, and one which corporate safety specialists or regulators may expect to see evidence for, is to consider the full range of possible effective control measures and eliminate those that are disproportionate in terms of combined effort versus benefit.

- Bow ties are graphical and they assist in communicating the ALARP concept to a wide range of staff from front line to senior management.

Thus, bow ties provide a qualitative tool to assist in ALARP justifications, better than simple judgment. However, if the area is contentious or very costly to address, then more quantitative tools such as LOPA or QRA may be necessary, with cost-benefit analysis, to demonstrate whether an additional barrier is 'justified' or 'grossly disproportionate' to the risk reduction that will be achieved.

7.4 OPERATIONALIZING BOW TIES (MOPO / SOOB)

Bow ties provide support for important operational uses such as for activity decisions (this section) and for incident investigation (next section).

During operations, there may be pressure to carry out critical tasks when certain barriers are known to be degraded or out of service. This can result in pressure on front-line personnel to carry out tasks they feel may be unsafe at that time. Stop work authority is designed to address this, but many staff are reluctant to employ this unless the indications of increased risk are very clear. The bow tie can help staff use their stop work authority to postpone the activity until the barriers are repaired or replaced with an equivalent temporary measure. This can be documented in a Manual of Permitted Operations (MOPO) or Summary of Operational Boundaries (SOOB) which is a vetted transparent decision-making tool. These terms are synonymous. An example of a MOPO application would be that if fire pump capacity was reduced due to a diesel pump being unavailable in a system with multiple pumps, then no hot work would be permitted until the fire water system was repaired or additional temporary firefighting resources were placed near the work.

Bow ties provide important assistance to the task of building MOPO / SOOB matrices as they have identified all the barriers related to relevant top events, including many important degradation controls.

A MOPO can be a very simple matrix listing all desired activities down one side and all documented barriers across the top. An experienced front-line manager / supervisor would help generate this list and define the barriers and degradation controls that must be functioning for each task. This might be done as a later stage design / construction task, before the facility is brought into operations, but after the operating philosophy and procedures are defined. A conceptual MOPO chart is shown in Table 7-1 applying to both the barriers and degradation controls on which they rely.

Table 7-1 shows that to carry out Task 1 then barrier A and degradation control B must be fully functional (i.e., not degraded). These barriers / degradation controls can be hardware, human or organizational. If during operations, degradation control B is known to be degraded and there is pressure to carry out Task 1, then front-line personnel can simply refer to the MOPO chart and postpone the activity until the degradation control is repaired or replaced with an equivalent temporary measure. This provides significant shielding to front-line personnel from undue pressure and helps assure senior management that production pressures are not leading to risk-taking at the workplace.

Table 7-1. Conceptual MOPO or SOOB Chart

Desired Task	Barrier A	Degradation Control B	Barrier C	. . .	Degradation Control N
Task 1	X	X			
Task 2	X		X		
. . .					
Task T	X	X			X

Note: X means required for that Task

A full example of a SOOB chart developed for the International Association of Drilling Contractors (IADC) is shown in Figure 7-2. The figure uses three separate matrices;

- Operations vs. Simultaneous Operations (SIMOPS)
  (e.g., crane operations cease during helicopter operations);

- Operations vs. Barriers with reduced effectiveness (MOPO)
  (e.g., well testing is prohibited if gas detectors are impaired);

- Operations vs. Operational risk factors (ORF)
  (e.g., high winds prevents crane operations).

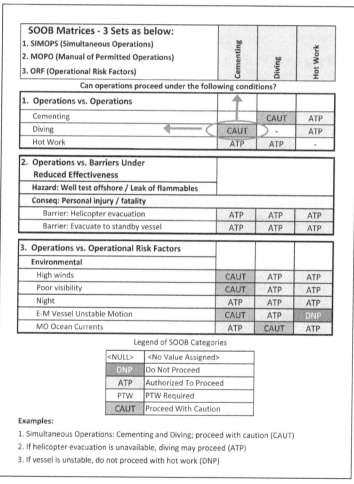

Figure 7-2. SOOB with SIMOPS, MOPO and Operational Risk Factors (IADC, 2015)

These three matrices can be used to assist site managers to determine, by reference to a prepared set of rules, what operations can or cannot be conducted (or what precautions must be in place) under certain conditions. The conditions may be categorized under three types;

1. Simultaneous operations which could increase risk;

2. Barriers that are degraded which may increase risk; or

3. Operational risk factors which may render an operation less safe.

Although not all of these relate to barriers, the concept includes consideration of barrier status, the understanding of which is enhanced in the bow tie context.

Another published MOPO implementation with full details of the three matrices approach for an oil and gas facility is illustrated on Figure 7-3.

The benefits of the MOPO approach include that it provides:

• Guidance for decision-making;

• A means of experience transfer from senior operational staff near retirement to the next generation of the workforce;

• A process to assess if additional controls are needed due to degraded barriers, external influences or simultaneous operations;

• An easily accessible set of comprehensive references;

• A consistent and accessible guidance across all assets of the same type;

• Assurance that corporate risk targets are being achieved even where barriers are degraded due to affected activities being postponed or barriers replaced; and

• A means to capture operational experience and to update the chart through the facility lifetime.

Many commercial bow tie software tools (see Appendix A) provide built-in MOPO / SOOB support. These charts can also be created in spreadsheet software using embedded links to direct the user to relevant documents or procedures (see Detman and Groot, 2011). Purely manual charts are also possible, although these can be unwieldy for large charts and make the task of updating the chart more difficult.

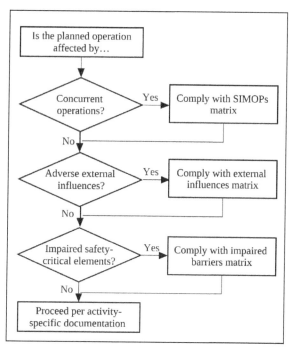

Figure 7-3. Operation of a MOPO System (Detman & Groot, 2011)

7.5 INCIDENT INVESTIGATION USING BOW TIES

Incident investigations should be carried out for all incidents and ideally for every significant near-miss. The aim of incident investigation is to identify the root causes, not just the immediate causes, to prevent recurrence. Investigation of incidents is a requirement in process safety regulations in the US, Europe, UK and Australia, among others. Incident investigation procedures are covered in many texts, including CCPS (2003) and EI (2015). These references cover several root cause analysis techniques. Root causes may be any combination of technical or HOF failures.

Representations of Swiss cheese models, particularly as part of incident investigations, have sometimes lined up twenty 'layers of cheese' that have failed. These investigations are not faulty, just that incident bow ties do not meet the barrier validity criteria of bow tie diagrams – as that is not the purpose of the investigation. These 'layers of cheese' may be components of a barrier, degradation controls (supporting barriers) or management system elements such as management of change.

The UKs Process Safety Leadership Group (PSLG, 2009) provided a detailed discussion, using the explosion and subsequent fires at the Buncefield oil storage facility in the UK in 2005 as the exemplar, of how reviewing an incident investigation can provide great insight into the role of people in barrier systems. The discussion also included consideration of how human barrier functions can be defeated both by engineering design and by organizational decisions. The PSLG also included many suggestions for the kinds of questions that investigators can ask to thoroughly probe how the human and organizational aspects of barrier systems failed in incidents.

Barrier and bow tie ideas are included in at least three established incident investigation methods; SOURCE (ABS, 2008), Tripod Beta (Energy Institute, 2015), and BSCAT (Pitblado et al, 2015). These are called 'barrier-based incident analysis' techniques. Simply put, whereas most incident analysis looks for why the incident occurred, they ask, "Why did the barriers fail which should have stopped the incident occurring?" Barrier and bow tie analyses graphically depict that, for an event to progress beyond the threat stage, one or more barriers must fail. However, in near-misses, some barriers in the pathway may work or all may fail though the ultimate consequence (an accident) does not occur (e.g., due to wind blowing the hazardous material away from exposed people). It is desirable for the investigation method to explicitly track specific barrier failures. McLeod (2015) provides a detailed discussion, including many prompts and questions that can be used in an incident investigation to explore in depth how human and organizational factors may have contributed to barrier failure.

The SOURCE method identifies Equipment Performance Gaps and Front-line Personnel Performance Gaps where the barriers have failed. This approach then works through the intermediate causes (the degradation factors) to identify the underlying root causes (i.e., the degradation controls that did not function as desired).

The Tripod Beta approach is a technique that requires both incident investigation expertise and good knowledge of human factors to properly apply the method. It uses an Agent (the Hazard), Object (the thing or person harmed by the agent), and Event (the incident) diagram to show the progression of events that occurred (what happened). Barriers that should have prevented the incident occurring are placed within the diagram (how the incident occurred). For each failed barrier, the investigator identifies an immediate cause (an operator error, for example), preconditions (the performance influencing factors that made the immediate cause more likely to happen), and underlying causes (the latent systemic failures; for example, decisions, policies, culture, that created or failed to control the preconditions) explaining why the incident happened. Because of the level of depth and thus effort required, Tripod Beta is often reserved for more serious incidents. However, some companies apply the method to small incidents as well, so scale is not a limitation. Tripod Beta bow ties are different to standard bow ties described in this book (see the upper part of Figure 7-4), but investigators can use a pre-existing bow tie to assist in the construction of the Agent-Object-Event diagram.

The BSCAT (Barrier Systematic Cause Analysis Technique) approach is intended to be applied by front-line supervisors and is applicable to all incidents. It combines the bow tie pathway for the incident with a traditional root cause tool (SCAT – Systematic Cause Analysis Technique) applied to each failed barrier in turn. SCAT defines, for each impaired barrier, its immediate cause, its root cause, and the management system action for improvement. BSCAT can use pre-existing bow ties as the basis for the incident bow tie (see the lower part of Figure 7-4).

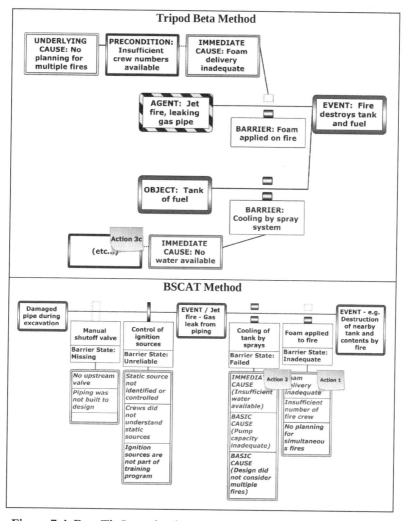

Figure 7-4. Bow Tie Investigation Methods (Tripod Beta and BSCAT)

The Tripod Beta approach creates the barrier diagram during the investigation and integrates human factors theory while BSCAT can use pre-existing facility bow ties, where these exist, and captures human and technical failures in the immediate and root causes using checklists. Generally, only one or two pathways of a pre-existing bow tie will apply to a specific accident, so the incident bow tie is simpler than a full all-threats bow tie. If a bow tie analysis exists, it is much easier for an incident investigator / analyst to immediately focus the investigation on the obvious failure pathways shown in the bow ties. Often the required level of detail is greater in an incident investigation than in a bow tie analysis, and this more detailed understanding can then be added to the existing bow tie after the investigation.

One important advantage of the barrier investigation approach is that it links incident investigation to the underlying barrier model for the facility (its bow tie) and the results of every investigation can reinforce workforce appreciation of the barrier model.

Incident / near miss information developed during the investigation should be used to update existing bow ties. This will improve the understanding of risk within the facility. Also, because pathways (whether prevention or consequence) are often reused in other bow ties, sometimes with little modification, these related pathways should be updated as well. This is a powerful means to share lessons learned across an organization.

The advantage of both investigation approaches is that they employ the barrier model as a driver for understanding the incident causation one barrier at a time, whereas older methods tend to collect all the causes together. Generally, the barrier failures occur for specific reasons and it is easier to understand the event causation if the root causes are directly mapped onto the failed / degraded barriers themselves rather than to the whole incident as a list.

7.6   REAL-TIME DASHBOARDS USING BOW TIES

Dashboards are a means to display the status of safety barriers in a high-level overview of overall facility safety status. The status of the barriers is typically presented in terms of dials or traffic light colors of green, yellow or red (functioning; degraded or failed; missing) but can be user customized. Other systems can be connected to automatically update the status of the barrier on the dashboard. While this is referred to as 'real time', the dashboard only shows the status from the time of the last update.

Such dashboards, however, can require a considerable effort to set up and maintain. It is therefore important to be clear on the purpose, target audience, and whether they supplement or replace parts of the existing barrier management program. Several companies are presently working on tools to monitor the health and status of safety barriers and have developed their own version of dashboards.

Scottish Power (Gray & Fenelon, 2013) has published a well-known example that uses a bow tie based dashboard system.

Benefits of bow tie based dashboards include:

- An 'at a glance' assessment of leading and lagging indicators for facility condition;
- Assurance that barriers are being maintained and monitored;
- Prompts for actions to restore degraded / failed barriers;
- Supports immediate decision-making using 'at-the-moment' information (e.g., barrier condition as an input to maintenance prioritization)
- Decision support at all levels in an organization from the frontline all the way up to senior executives; and
- Wide sharing of information to prevent repeat incidents.

Although 'real time' bow tie dashboards bring benefits, there are also difficulties in their creation and use:

- Many barriers do not lend themselves to online monitoring;
- The health indication of the barrier is only as good as the level of detail that supports the 'online' assessment;
- Some barrier status indicators can be outdated and not truly capture current conditions (if based on an audit, for example);
- The online system can remove or reduce the human input and analysis of barrier condition; and
- The dashboard may not normally differentiate which degraded barriers are critical and therefore which should receive priority.

Methods for dashboards are still under development. Given the diversity of industries, safety management systems, and decision-making needs, no single approach is likely to emerge as industry consensus.

7.7   BARRIER AND DEGRADATION CONTROL VERIFICATION

Barrier and degradation controls verification is like an audit. Audits along with metrics are the checking part of the management system loop: plan – do – check – act. They systematically address each element of the management system to confirm its proper operation. However, in the context of bow tie diagrams, there are meaningful activities, termed audits, which address issues that are more technical. Thus, rather than addressing management system elements, the barrier audit reviews the functionality and health of barriers displayed on the bow tie.

Verification is normally conducted at the end of construction to confirm all barriers are installed as defined in the approved design and in PHA or other safety studies; this is also known as a readiness or pre-startup safety review. Verification is also conducted through life to confirm that barriers continue to function as intended or that changes approved by MOC processes are implemented.

Bow tie diagrams can be used in a 'live' sense to show the condition of barriers. This fits well with the UK system for safety cases with defined performance standards for safety critical elements (typically hardware barriers). Each of these is supported by an inspection, testing and maintenance strategy (called a written scheme) that is designed to keep that barrier at its performance standard through life. The barrier or degradation control condition would be displayed as metadata using functionality available in most bow tie software. In a pre-verification mode, this might be to define the means that would be used for verification; for example, test the flow rate and pressure of fire pumps, verify the performance of the work permit system by past permit document review and by task observation, etc. In a post-verification mode, the metadata can be used to display the results. These two verification modes are summarized on Figure 7-5.

By displaying this information on the bow tie, all staff know how the barrier is to be verified and the status from the most recent verification. The date can be added as a note if this is helpful.

Statoil has developed a very thorough barrier audit program that has the same rigor as a detailed management system audit. The Statoil program is described by Kortner, et al (2001). Through this procedure, the status of all key technical barriers in terms of their design, their condition, and their operation are reviewed. In the design review, the initial design basis is compared to the current demands in case these have changed. The condition is a direct indication of the status of the barrier, and the operation addresses the training and competence of staff responsible for the barrier. A total score is assigned to each barrier and where this falls below nominal performance then remedial action is required. The program was developed so that the Statoil CEO could periodically assure the regulator that their facilities remain in a safe operational condition. This written affirmation was intended to link the CEO personally to the facility safety. The barrier status information can be displayed on bow tie diagrams using the format described above.

7.8   BOW TIE CHAINING

Bow tie chaining or linking is an advanced feature of the diagram used to display multiple top events in an event sequence. This can be useful for complex failure modes, or just representing end-to-end consideration of risks on a 'horizontal' plane (e.g., all bow ties are in the operations environment). Additionally, they can be used for 'rolling up' bow ties from operations 'vertically' to show the impact from operations to issues at management levels.

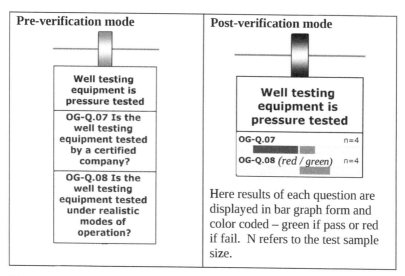

Figure 7-5. Displaying Verification Information on Bow Tie Diagrams

Chaining can be conducted (especially in some bow tie software) by linking a threat or a consequence from one bow tie to a top event of another bow tie. Thus, a top event in one diagram can also become a threat in the next. Chaining can also be done manually.

The starting point may be constructing a simple event relationship analysis to plan the required bow ties. This is performed by listing each of the major events (starting with fire / explosion at the top of the page) that may occur at a facility, and drawing arrows to show what can cause other listed events. Some of the events will clearly be top events, and as the arrows indicate, some will be candidates for threats on one bow tie and top events on another, etc.

An example of 'horizontal' chaining could be structural collapse within an offshore production platform or ship collision. Both have immediate consequences such as loss of life but they can also be a cause of loss of containment and a hydrocarbon fire and explosion, which appears in the chained bow tie.

An example of 'vertical' chaining would be as follows. A top event for a bow tie in the boardroom may read 'failure to achieve corporate goals'. Each of the threats may describe major risks around finance, human resources, major process safety or personal safety incidents, or 'chemical product not delivered to agreed contract terms'. In the sales department's bow tie, one level down, a bow tie top event reads 'Product not delivered to agreed contract terms', and one of several threats to this is 'Inventory not available when needed'. One level further down, this might read as the top event for the inventory department's bow tie, but one of their threats reads 'Production failure in petrochemicals facility'.

Care should be taken not to push such a 'chained' bow tie towards a specific preferred event sequence, as opposed to its basic function as a general risk management model. In addition, it may add complexity to a diagram and novice users may not understand the format, thereby defeating the key communication objective.

Chaining can also be accomplished by merely placing text on threats, consequences or on main pathway barriers to refer to other diagrams.

Advantages of chaining include:

• Demonstrates end-to-end consideration and management of risks;

• Avoids overlap (two bow ties covering some of the same risks);

• Helps to structure and phrase the right top events, direct threats and consequences, and promotes adherence to the bow tie methodology;

• Avoids repetition (and therefore double-management) of barriers on 'horizontal' and 'vertical' bow ties. It does this because (for example) on a threat, the author can reference and invoke the barriers on another bow tie ('See bow tie 04');

• Vertical chaining can give managers right up to the Board level, a view of holistic risk management of major risks in the organization; and

• It is a step toward recognizing, and modeling within the bow tie context, the many interactions that can occur in complex and tightly coupled systems.

Disadvantages of chaining include:

• Novice readers may be confused;

• Care is needed to add barriers that address the new top event and not just assume that barriers invoked on the preceding threat are sufficient; and

• Novice analysts may pursue the wrong models for the sake of completing the 'jigsaw' – i.e., forcing a bow tie into a space that is neither needed nor logical.

7.9   ENTERPRISE-WIDE ANALYSIS AND WINDOW ON SYSTEMIC RISKS

Bow ties can serve useful functions at the enterprise level. The idea of corporate bow tie templates was discussed in Section 6.3 and how these can provide guidance to individual facilities on how MAE risks might be well managed. This is a top-down application.

The bottom-up application would collect all the facility level bow ties and compare these for similar threats. It may be that individual facilities have developed innovative solutions that are superior to the nominal corporate good practice bow tie templates and these should be updated accordingly. Similarly, information on verification activities can be compared to identify if certain barriers are experiencing degradation issues at many locations.

An interesting option is to document barrier failures in near-miss events and actual incidents – this is a lessons-learned application. It is a feature of bow ties that many barriers are repeated on multiple pathways and in multiple bow ties. Thus, there are likely to be multiple demands on the same barrier type at many facilities (e.g. Lock-out Tag-out barriers, ignition control barriers, etc.). If a barrier-based incident analysis is conducted, then it is relatively easy to combine all degradations or failures of barriers in incidents and to search for common immediate or root causes or actions for improvement. This can identify enterprise-wide risk and barrier issues and help in ongoing improvement activities. One advantage of the bow tie approach is that lessons for specific barriers can be communicated to owners of those specific barriers at every facility and an email generated automatically. It does not require special insight from the corporate safety group to identify who specifically should be informed. Barrier owners might collect such emails and deal with them periodically to avoid being overwhelmed with email actions.

7.10   CONCLUSIONS

This chapter has summarized several advanced uses of bow ties. These include linking bow ties to other risk approaches, using bow ties for ALARP demonstrations, MOPO / SOOB application for activity approvals when some barriers are degraded, barrier based incident investigations, real-time dashboards, barrier verification, chaining and enterprise lessons.

These uses can provide useful and powerful extensions to the basic bow tie objective, which might be seen fundamentally as risk communication. As bow ties become more widely used in industry, it is likely there will be further uses developed or refinements on the ones presented here.

APPENDIX A – SOFTWARE TOOLS

SOFTWARE USED FOR BOW TIE DIAGRAMS

Specialist software packages can make it easier or quicker to construct bow ties. Some software packages may offer added benefits as a means to link barriers to other systems such as performance indicators or accident investigation outputs. However, bow ties do not need to be constructed with specialist software and can simply be constructed using pen and paper, sticky notes, or common software used for creating spreadsheet presentations and flow diagrams.

Table A-1 summarizes a selection of better-known current software tools used to construct bow tie diagrams (listed in alphabetical order). Web addresses were current at the time of publishing.

Table A-1. Listing of Main Features of Commercial Bow Tie Software

Software and Developer	Description
BowTie Pro BowTie Pro Ltd www.bowtiepro.com	This software is developed by Bowtie Pro Ltd of Aberdeen. It offers a full range of bow tie features. It has an easy to use GUI and can display a wide range of additional barrier metadata. The tool is supported by a user manual and training courses. It is available in a standalone or networked version.
BowTieXP & IncidentXP CGE Risk Management Solutions, Netherlands www.cgerisk.com	BowTieXP is a full capability bow tie tool developed by CGE Risk Management Solutions in the Netherlands. As well as providing standard bow tie functionality, it allows entry of a wide range of barrier metadata including owner, reliability, effectiveness, status, outstanding actions, etc., in order to provide advanced barrier management functionality to individuals or the enterprise management team. The GUI allows different levels of detail to be displayed and color-coding and notes provide extra information aiding communication. The tool supports SOOB charts (SIMOPS, MOPO and Operational Risk Factor) to aid decision-making when some barriers are degraded or certain conditions are intolerable. Multiple language versions are available. The tool can be run on a standalone basis or in an enterprise version. The extension IncidentXP implements several methods including: Tripod Beta, BSCAT, TOP SET®, and Root Cause Analysis. When developing BSCAT charts, pre-existing Bow Tie XP diagrams can be imported as a basis for the incident bow tie build-out. Incident XP provides a timeline tool, and graphically displays all barrier performance during the incident with individual barrier Root Cause Analysis. Barrier failure information can be displayed back onto the original full set of bow ties. Additional extensions provide hazard and incident registers, provide LOPA and quantitative extensions, and auditing and compliance functions. AuditXP allows customized questions to be attached to each barrier for periodic audit surveys. Red / green results are visible on the risk barriers, alongside the incident analysis results if so desired. The BowTieXP and BSCAT and other software tools are supported by full documentation and training and the software is regularly updated.

Table A-1. Listing of Main Features of Commercial Bow Tie Software, continued

Software and Developer	Description
Graphic Software packages Microsoft, Adobe, others.	The bow tie approach can be implemented in a wide range of graphic software – where the aim is communication of barriers rather than creating a risk management database with multiple additional metadata. While it might be possible to use simple graphic tools such as PowerPoint, it is better handled in tools that provide layers such as Visio, Paint.net, and Adobe Creative Suite. Vector graphics are superior to bitmap software as edits are much easier for future changes. The bow tie packages listed in this appendix can export figures into the general-purpose packages for additional customization. These tools are not optimized for bow tie applications and thus updating diagrams and storing related metadata can be difficult.
Synergi DNV GL Software, Norway www.dnvgl.com/software	This tool, originally released as EASYRISK for project risk management, was extended to full risk management capability when ported into Synergi. The tool provides standard bow tie functionality and its unique feature is its linkage to the full incident database maintained by many companies in Synergi. The bow tie feature extends the risk management capabilities of Synergi rather than being the primary purpose of the tool. As with the other software, there are user manuals, training courses, and regular software updates.
THESIS ABS Group www.abs-group.com	This software was developed by ABS Group initially for Shell to help implementation of the HEMP system (Hazards and Effects Management Process). The software is now commercially available. It offers a full range of bow tie capabilities as well as related safety functions. The functional diagram for the tool includes bow ties, hazards and effects register, LOPA, activities / tasks, and shortfalls / remedial action plans. The software is designed with a user-friendly GUI and can be run either as a standalone package or on a full corporate network. THESIS has been aligned with this concept book using updated terminology including the degradation factor and degradation control terms, and metadata descriptors. THESIS creates a risk management database beneath the bow tie diagram and this holds a wide range of risk management information – including barrier lists, barrier owners, outstanding actions, etc. The tool is supported by user manuals and training courses and the software is regularly updated.

APPENDIX B – CASE STUDY

INTRODUCTION

The purpose of this Appendix is to demonstrate how to build a bow tie following the methods presented earlier in this book. The bow tie follows the guidance in this book, but does not necessarily represent current best engineering practice – that would need to be determined by a team for a specific design in a specific location. The example presented is a bow tie for volatile hydrocarbons under pressure in a pipeline. The emphasis in this example is on technical threats, while Appendix C addresses human factors in more detail for a different example.

VOLATILE HYDROCARBONS UNDER PRESSURE IN A PIPELINE

A risk assessment has been carried out for a pipeline under pressure and since loss of containment is a key safety and environmental issue, it was decided that a bow tie should be created for a loss of containment event.

Recall the steps defined in Figure 3-1 for bow tie construction (repeated below).

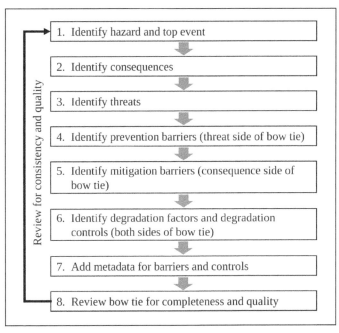

Figure B-1. Bow Tie Creation Flow Chart

Step 1) The Hazard and Top Event

The management team decides that the scope for the bow tie is to identify the primary barriers only, and not to document all the degradation pathways. This will keep the diagram simple and focus on the most important barriers. The hazard here is 'Volatile Hydrocarbons under pressure in pipeline.' It would be incorrect to nominate the leak as the hazard as discussed in Section 2.2.2. It is the top event, which would be the loss of containment of the hydrocarbon.

Step 2) Identify Consequences

The primary consequences here are injuries to people from fire and / or explosion, and damage to the asset from the same fire / explosion, and spill to the environment. There can be other consequences such as product loss, reputation, legal proceedings, etc., but these are either less important or do not introduce new important barriers. The diagram is limited to the three primary consequences.

Step 3) Identify Threats

The team initially identified many separate threats. These included: rupture of the pipeline due to overpressure from temperature excursion, or external corrosion of pipeline or overpressure due to ESD downstream. External impact was also an identified concern. There may be other threats related to this example but only a few are shown to highlight the concept.

The first step is to see if any important threats have been omitted from the assessment. A review of historical pipeline data shows that internal corrosion is a major threat to pipelines as well as fatigue so these are added. Internal and external corrosion might be a candidate to combine – but the barriers are quite different so it is decided to maintain these as separate threats. In the example below, internal corrosion and impact are not shown to save on space.

Step 4) Identify Prevention Barriers

For the first threat: rupture due to external corrosion of pipeline, the team identified several barriers:

1) Use material to suit the process fluid and the environment;
2) Pipeline route selection;
3) Cathodic protection system;
4) Anti-corrosion coating; and
5) Periodic direct inspection (as required by regulation).

Barriers must have the capability on their own to prevent or mitigate a bow tie sequence and meet all the requirements for a barrier – effective, independent, and auditable. The first barrier 'use of material to suit process and environment' meets the requirements for a barrier but is not a valid barrier (e.g., containing the detect-decide-act elements). The team recognizes this and the barrier is renamed to the passive barrier 'Steel containment envelope' with the possibility of the wrong material being chosen, so a degradation factor control of 'design and material selection for compatibility with the environment.' The second barrier 'Pipeline route selection' is better termed as a degradation control as it cannot on its own prevent the top event. Thus, this should be handled either in a degradation factor or combined with the first barrier. Similarly, counting the two corrosion protections as separate barriers is incorrect and these should be counted as one barrier: 'External corrosion protection system' (for simplicity, the additional words 'coating and cathodic protection' are captured as an extra description). The degradation of these can be handled as two separate degradation pathways as the degradation control is different for these two systems. Periodic direct inspection of the pipeline is used to ensure that the materials of the pipeline remain fit and valid. As such, it is a degradation control against the factory coatings being compromised due to local jointing and modifications. Thus, the original list of five barriers is converted into two barriers that meet the requirement for independence, effectiveness, and auditability:

- Steel containment envelope;
- External corrosion protection system.

A similar exercise is carried out for the other threats.

Step 5) Identify Mitigation Barriers

For the consequence: Fire and explosion, the original risk assessment documents seven barriers:

- Leak detection system;
- Emergency shutdown isolation valves;
- Periodic aerial observation (vegetation damage indicative of leak);
- Ignition control;
- Evacuation of vulnerable people;
- Use of fire protection equipment (cooling water, foam, etc.); and
- Extinguish fire (after appropriate safety review).

Not all these barriers meet the tests outlined in Chapter 2.

A leak detection system does not meet the three parts of an active control: detect – decide – act. Therefore, the first two barriers and the unstated operator response become the single barrier 'Leak detection, operator assessment and ESD isolation'. In the diagram, for brevity, this might be shortened to 'Leak detection and manual isolation', but it does include all three aspects.

The periodic aerial observation barrier is useful to identify smaller leaks below the threshold of the leak detection system. It can be included in the first barrier. Major leaks are detected by the leak detection system and small leaks by aerial observation. It still requires operator response and isolation to complete the barrier. Periodic aerial observation is also a degradation control against external impact on the pipeline to give early warning of, for example, construction equipment being prepared for excavation activities near the pipeline.

Ignition control is not present for most of the pipeline route except in the plant areas at the source and destination. Since these areas would be treated by separate bow ties, it is not helpful to add a barrier that mostly does not apply.

'Execution of fire emergency response plan' is a good title for a barrier which might also include mobilizing local emergency responders, establishing an emergency perimeter, and notifying regional authorities. Use of fire protection equipment and extinguishing the fire are part of a single barrier which might be called 'active fire-fighting system'. Thus, the mitigation barriers against the consequence of fire and explosion might be:

- Leak detection and manual isolation;
- Execution of fire emergency response plan; and
- Active firefighting system.

Evacuation of vulnerable people is part of an emergency response. While each of these barriers arguably can reduce or prevent harm to people, the barriers against the consequence of injury are limited to those which specifically address the personal isolation, protection and recovery aspect (exclusion zone, evacuation and medical response). Otherwise, each consequence line can unnecessarily become repetitive with the incremental value of barriers such as 'firefighting system'.

For the environment consequence, only the first barrier is effective at directly mitigating the consequence of spill to the environment, and is joined by a clean-up program.

Step 6) Identify Degradation Factors and Controls

In principle, all barriers have systems that are employed to maintain their effectiveness. This includes operator training, inspection and maintenance, etc. It is not helpful to repeat these everywhere. In this example, the 'external corrosion protection system' would be a good candidate to show two separate degradation pathways – for external coating failure from local jointing work and cathodic protection failure. The barrier 'manual response to pressure switch alarm' could fail with incorrect set points. Similarly, on the mitigation side, 'Leak detection and isolation' might be expanded to show the difficulties in achieving this in remote areas.

In this example, a degradation pathway is shown for the barrier 'external corrosion protection system' on the threat side. A degradation pathway is shown for 'evacuation of personnel' on the consequence side as well as 'leak detection and manual isolation of inventories'.

The same process can be applied to all the arms (threat and consequence), but for limitations of space only a few are shown here.

Step 7) Add Metadata for Barriers and Degradation Controls

This step captures knowledge of barrier / degradation control owners, effectiveness, performance standard, status, etc., using the knowledge available from the team.

In this example metadata was added to the barrier 'PSV (Relief valves, rupture discs and pins)' as below:

- Barrier type: Passive hardware;
- Barrier owner: Engineering manager; and
- Criticality: Critical.

The same is done for any and all barriers / degradation controls according to the scope of the workshop. Metadata is not displayed on the bow tie due to space limitations.

Step 8) Review Bow Tie

The bow tie is reviewed for quality and consistency. The bow tie presented here is to demonstrate consistency with the guidance in this book. It is not necessarily representative of current good engineering practices. The reader could use the following questions one through seven to critically review the bow tie presented in Figure B-2 and Figure B-3 to see if it represents current good practices.

1. Has the bow tie been developed to address all the major accident events identified in prior PHA studies and does it match the study terms of reference?
2. Did the bow tie workshop team include the right mix of people, was sufficient time permitted to allow a thorough analysis, and has the quality review been carried out by an independent specialist?
3. Do all elements of the bow tie match the rule set agreed (i.e., hazard, top event, threat, consequence, barrier, degradation factor, degradation control as summarized in Chapter 2)?
4. Is the terminology used correct and consistent across all bow ties?
5. Does the bow tie contain any structural errors (e.g., degradation controls on main pathway, or includes ineffective barriers)?

6. Where multiple bow ties have been created, are they mutually consistent (similar top events call on similar barriers, and barrier names, degradation factors, degradation controls, are consistent with other appearances, etc.)?

7. Does the complexity and content of the bow tie match the intended audiences?

Refer to Section 3.3 for a full list of questions and criteria. This list includes specific checks on the hazard and top event, threats and consequences, barriers and degradation controls.

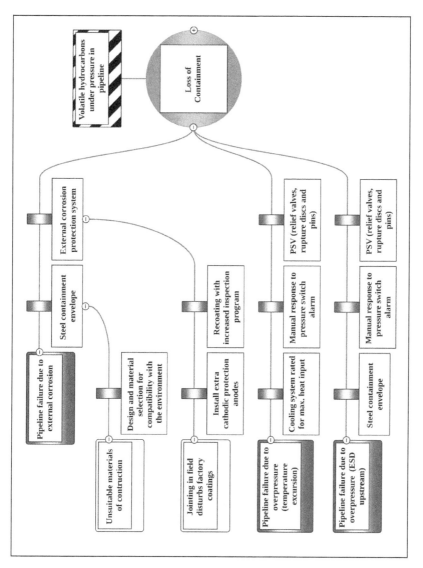

Figure B-2. Pipeline Loss of Containment Bow Tie – Threat Leg

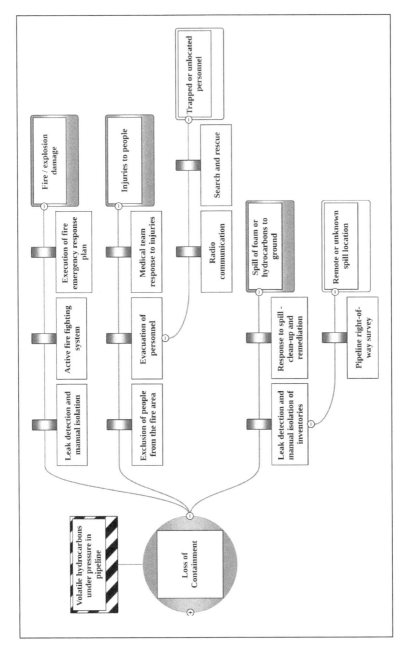

Figure B-3. Pipeline Loss of Containment Bow Tie – Consequence Leg

APPENDIX C – Multi-Level Bow Ties

MULTI-LEVEL BOW TIE FOR TANK OVERFILL

The concept of multi-level bow ties was introduced in Chapter 4 with particular reference to their ability to enhance analysis for human factors issues. These typically involve a larger number of degradation controls than barriers, and these appear in degradation factor build-outs. A concept figure was included in Chapter 4, Figure 4-5, but a worked example is helpful to understand the concept better.

The human factors sub-committee for this book developed an example for a tank overfill top event. This event was inspired by the Buncefield accident in the UK, but the detail is different and selected to show better the idea of multi-level bow ties.

The bow tie attempts to match the level of detail required for a real bow tie analysis and thus it is reasonably complex. Nonetheless, it is not complete. For example, no degradation factors are shown for the 'fill plan and proactive monitoring' barrier. CCPS maintains a website for this concept book where the full bow tie created in software is available (see preface Page xxiii).

The example is in three parts and is taken from Manton et al (2017). Figure C-1 shows a summary bow tie for one threat (Fill volume exceeds tank ullage), Figure C-2 shows the bow tie expansion to show the main pathways and barriers on the threat leg and Figure C-3 shows the consequence leg. For those not familiar with the threat, the planned fill volume exceeds the allowable empty space above the initial tank level. For simplicity, the consequence side of the bow tie is not built-out for this example.

Note: It may appear to some that the first two barriers in Figure C-2 are not independent, as both appear to rely on the same operator. This was in fact reviewed by the UK working party (UKPIA) addressing the follow-up to Buncefield and they concluded that these two barriers were in fact independent as they were separated significantly in time and that a single operator failure (e.g., temporary absence) would not defeat both.

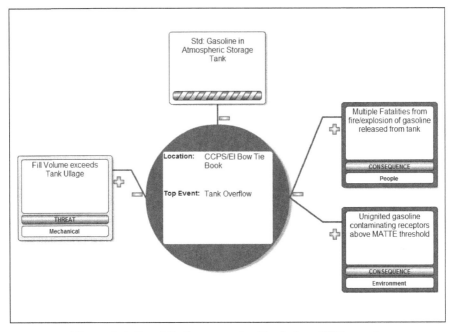

Figure C-1. Bow Tie Summary for Tank Overfill Example

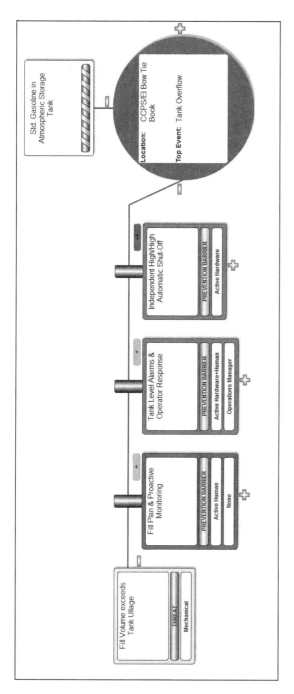

Figure C-2. Bow Tie Expansion Showing Main Pathways Only (Prevention Side)

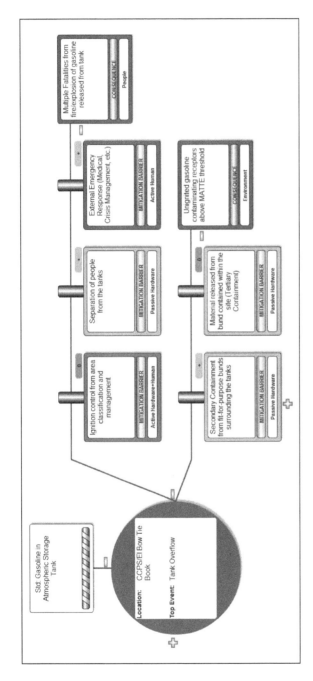

Figure C-3. Bow Tie Expansion Showing Main Pathways Only (Consequence Side)

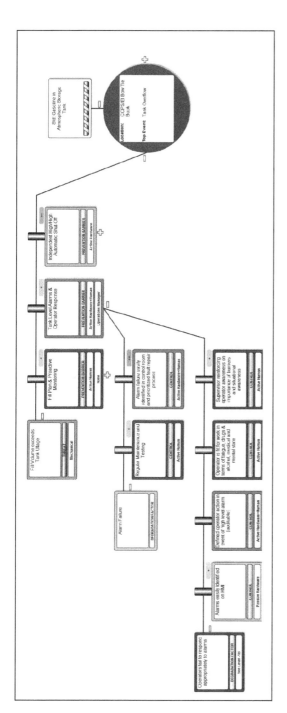

Figure C-4. Standard Bow Tie Showing Main Pathways and Degradation Factors (Prevention Side)

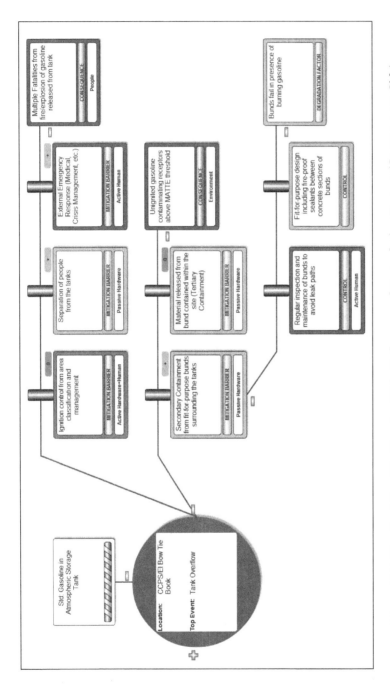

Figure C-5. Standard Bow Tie Showing Main Pathways and Degradation Factors (Consequence Side)

An explanation of Figure C-4 and Figure C-5 is given in Table C-1.

The Level 1 Extension bow tie is shown developed for one degradation factor: 'Operators fail to respond appropriately to alarms.' Examination of Figure C-2 and Figure C-3 would allow that each of the two degradation controls can be extended with deeper level degradation controls. The contents allow display of ideas from Hudson and Hudson (2015) on presenting regulatory factors into the bow tie (e.g., stop work requirements addressed here by the STOP culture degradation control).

The left side of Figure C-6 shows degradation factors and degradation controls to ensure the operator can see the alarm, and the right side shows degradation factors and degradation controls to ensure the operator knows how to respond to the alarm. Manton et al (2017) provides additional examples.

The Level 1 Extension shows new information which is not easily displayed on the standard bow tie due to increasing complexity of the figure. Each degradation control along the main pathway is shown as having its own degradation pathway and two are displayed here. In principle, there could be multiple degradation factors for each degradation control, but it is useful to try to limit the complexity being added on this extension level.

Taking the first degradation control (left side of Figure C-6): 'Alarms easily identified on HMI', the Level 1 Extension shows two degradation factors supporting this control: 'Alarm overload' and 'Alarms hidden behind other

Table C-1. Explanation of Standard Bow Tie Figure C-4 and Figure C-5

Bow Tie Element	Description
Hazard	Gasoline in atmospheric storage tank
Top event	Tank overflow
Threat	Fill volume exceeds tank ullage (the remaining available space in the tank)
Barriers	1) Fill plan and proactive monitoring of plan 2) Tank level alarms and operator response 3) Independent High-High automatic shut-off
Degradation factors	Shown here for Prevention-side Barrier 2 only Barrier 2: two degradation factors (alarm failure and operator failure), showing two and four degradation controls, respectively Shown here for Mitigation-side Consequence 2 (environment impact MATTE: Major Accident to the Environment) Barrier 1: one degradation factor (bund failure) with two degradation controls

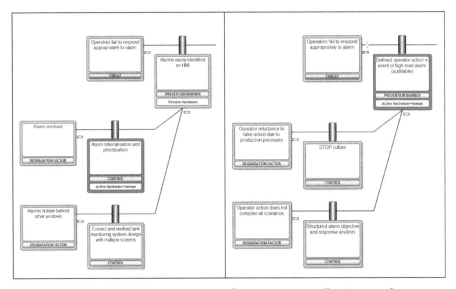

Figure C-6. Level 1 Extension Bow Tie for Operators Fail to Respond Degradation Factor (for Degradation Controls 1 and 2)

windows'. These are important degradation controls, but they do not show in standard bow ties and this is the value of Level 1 Extension. Another advantage is that these are more general processes, not only linked to the specific barriers being examined here, and they may be applied to other bow ties without change. Wherever there is a degradation control 'Alarms easily identified on HMI' in any other bow tie, then this Level 1 Extension set of degradation controls will be applicable, assuming the facility is consistent in its HMI design philosophy.

Thus, while it appears that Level 1 Extension makes the bow tie more complicated, it does show important degradation controls not evident in the standard bow tie and it avoids duplication for these more general degradation controls.

The full build-out of this bow tie would appear as shown in Figure C-7. This Level 1 Extension, which is only for one threat pathway, would be too complex to communicate to operational staff for typical bow ties with five to eight threat pathways. It is more the domain of subject matter experts, process safety specialists, and senior managers who need to understand how degradation controls are themselves supported or where controls required by regulation fit into the wider perspective of barriers and degradation controls.

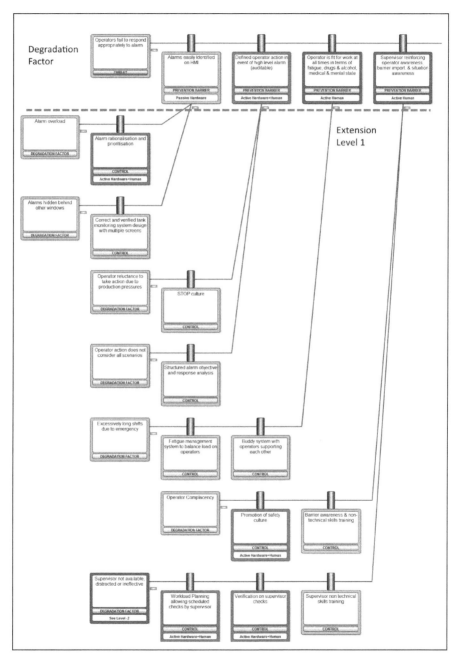

Figure C-7. Standard Bow Tie Degradation Pathway and Level 1 Extension showing Degradation Controls supporting Higher Degradation Controls

REFERENCES

ABS (2008) *Root Cause Analysis Handbook: A Guide to Efficient and Effective Incident Investigation* (Third Edition) by Vanden Heuvel, Lorenzo, Jackson. ABS Consulting.

API (2016) *Process Safety Performance Indicators for the Refining and Petrochemical Industries*, API RP 754, 2nd Ed., American Petroleum Institute, Washington DC.

CCPS (2000) *Guidelines for Chemical Process Quantitative Risk Analysis.* 2nd Ed. American Institute of Chemical Engineers, John Wiley & Sons, New Jersey.

CCPS (2001) *Layer of Protection Analysis: Simplified Process Risk Assessment.* American Institute of Chemical Engineers, John Wiley & Sons, New Jersey.

CCPS (2003) *Guidelines for Investigating Chemical Process Incidents, 2nd Edition.* American Institute of Chemical Engineers, John Wiley & Sons, New Jersey.

CCPS (2006) *Guidelines for Mechanical Integrity Systems,* American Institute of Chemical Engineers, John Wiley & Sons, New Jersey ork.

CCPS (2007) *Guidelines for Risk Based Process Safey.* American Institute of Chemical Engineers, John Wiley & Sons, New Jersey.

CCPS (2008a) *Guidelines for Hazard Evaluation Procedures, 3rd Ed.* American Institute of Chemical Engineers, John Wiley & Sons, New Jersey.

CCPS (2008b) *Guidelines for the Management of Change for Process Safety.* American Institute of Chemical Engineers, John Wiley & Sons, New Jersey.

CCPS (2008c) *Incidents that define process safety.* American Institute of Chemical Engineers, John Wiley & Sons, New Jersey.

CCPS. (2009) *Guidelines for Developing Quantitative Safety Criteria.* American Institute of Chemical Engineers, John Wiley & Sons, New Jersey.

CCPS (2010) *Guidelines for Process Safety Metrics,* American Institute of Chemical Engineers, John Wiley & Sons, New Jersey.

CCPS (2011) *Conduct of Operations and Operational Discipline: For Improving Process Safety in Industry.* American Institute of Chemical Engineers, John Wiley & Sons, New Jersey.

CCPS (2013a) *Guidelines for Managing Process Safety Risks During Organizational Change.* American Institute of Chemical Engineers, John Wiley & Sons, New Jersey.

CCPS (2013b) *Process Safety Leading Indicators Industry Survey*. American Institute of Chemical Engineers, John Wiley & Sons, New Jersey.

CCPS (2013c) *Guidelines for Enabling Conditions and Conditional Modifiers in Layer of Protection Analysis*. American Institute of Chemical Engineers, John Wiley & Sons, New Jersey.

CCPS (2015) *Guidelines for Initiating Events and Independent Protection Layers in Layer Protection Analysis*. American Institute of Chemical Engineers, John Wiley & Sons, New Jersey.

CIEHF (2016) *Human Factors in Barrier Management*, White Paper, Chartered Institute of Ergonomics and Human Factors. (Website checked March 2017)

CSB (2016) *Drilling Rig Explosion and Fire at the Macondo Well*, Chemical Safety Board Investigation Report, Volume 3 Report No. 2010-10-I-OS, issued 12 Apr 2016, Washington.

Detman, D., & Groot, G. (2011) *Shell's Experience Implementing a Manual of Permitted Operations. 14th Annual Symposium Mary Kay O'Connor Process Safety Center*. College Station: Texas A&M University.

Energy Institute (2010) *High level framework for process safety management* (known as 'PSM Framework'). EI website (checked Feb 14, 2018) (https://publishing.energyinst.org/topics/process-safety/leadership/high-level-framework-for-process-safety-management)

Energy Institute (2007) *Guidelines for the management of safety critical elements*, ISBN: 9780852934623, 2nd edition.

Energy Institute (2010) *Human factors performance indicators for the energy and related process industries*, 1st Edition, ISBN: 9780852935873.

Energy Institute (2012) *Guidelines for the identification and management of environmentally critical elements*, 1st Edition, ISBN: 978 0 85293 632 0.

Energy Institute. (2015) *Tripod Beta: Guidance on using Tripod Beta in the Investigation and Analysis of Incidents, Accidents and Business Losses*.

Energy Institute. (2016) *Understanding your culture, Hearts and Minds Toolkit*. Retrieved Dec 19, 2016, from http://publishing.energyinst.org/

Energy Institute (2016) *Managing Rule Breaking, Hearts and Minds Toolkit*. Retrieved Dec 19, 2016, from http://publishing.energyinst.org/

Energy Institute (2016) *Learning from incidents, accidents and events*, 1st edition, ISBN: 9780852939277.

Gray, D., & Fenelon, E. (2013) *The Implementation of Effective Key Performance Indicators to Manage Major Hazard Risks at Scottish Power*. Piper25. Aberdeen.

Hollnagel, E. (2012) *FRAM: the Functional Resonance Analysis Method.* Ashgate.

Hollnagel, E., & Leonhardt, J. (2013) *From Safety-I to Safety-II: A White Paper.* European Organization for the Safety of Air Navigation (EUROCONTROL).

Hollnagel, Erik (2014) *Safety-I and Safety-II: The Past and Future of Safety Management,* CRC Press; 1 edition.

Hopkins, A. (2000) *Lessons from Longford: The Esso Gas Plant Explosion.* CCH Australia.

HSE (1995) *Offshore Installations (Prevention of Fire and Explosion, and Emergency Response) Regulations 1995.* UK National Archives. Retrieved from Legislation.gov.uk: http://www.legislation.gov.uk/uksi/1995/743/contents/made

HSE (2001) *Reducing Risk and Protecting People.* UK Health and Safety Executive.

HSE (2006) *Developing Process Safety Indicators: A Step-by-Step Guide for Chemical and Major Hazard Industries. HSG 254.* UK Health and Safety Executive.

HSE (2008) *Optimising Hazard Management by Workforce Engagement and Supervision,* UK Health and Safety Executive. Research Report rr637 - http://www.hse.gov.uk/research/rrhtm/rr637.htm

HSE (2013a) *HID Regulatory Model, Safety Management in Major Hazard Industries.* UK Health and Safety Executive.

HSE (2013b) *Managing for Health and Safety.* HSG65. UK Health and Safety Executive.

Hudson, P. (2014) *Accident causation models, management and the law,* Journal of Risk Research.

Hudson, P., & Hudson, T. (2015). *Integrating Cultural and Regulatory Factors in the Bowtie: Moving from Hand-Waving to Rigor,* In: Ebrahimipour V., Yacout S. (eds) Ontology Modeling in Physical Asset Integrity Management. Springer, Cham.

IADC (2015). *Health Safety and Environmental Case Guidelines for Mobile Offshore Drilling Units.* Issue 3.6, International Association of Drilling Contractors, Houston.

IAEA (1999). *Basic Safety Principles for Nuclear Power Plants: 75-INSAG-3 Rev. 1.* Vienna: The International Atomic Energy Agency.

ICCA (2016) *Guidance for reporting on the ICCA globally harmonized process safety metric,* International Council of Chemical Association.

IChemE Safety Centre (2015) Lead Process Safety Metrics – selecting, tracking, and learning, Melbourne.

IOGP (2015) *Safety performance indicators – 2014 data*, The International Association of Oil and Gas Producers, http://www.iogp.org/pubs/2014s.pdf

IOGP (2011, November) *Process Safety - Recommended Practice on Key Performance Indicators*. Report No. 456. The International Association of Oil and Gas Producers.

IOGP (2016) *Process Safety – Leading key performance indicators Supplement to Report No. 456*. Report No. 556.

ISO 17776 (2016) *Petroleum and natural gas industries -- Offshore production installations -- Major accident hazard management during the design of new installations*. International Standards Organization. Switzerland.

ISO 31000 (2009) Risk management – Principles and guidelines. International Standards Organization, Switzerland.

Kanki, B. G., Helmreich, R. L., and Anca, J., (2010) *Crew Resource Management* Academic Press/Elsevier, 2nd Ed.

Kortner, H., Sorum, M., & Brandstorp, J. (2001) *Framework for Life-Cycle Assessment of Technical Safety conditions at Statoil Operated Plants*. ESReDA Seminar on Lifetime Management, November 5-6. Erlangen.

Leveson, N (2011) *Engineering a safer world: Systems thinking applied to safety*. MIT Press.

Manton M., Johnson M., Pitblado R., Cowley C., McGrath T., McLeod R., Miles R. (2017) *Representing Human Factors in Bow Ties as per the new CCPS/EI Book*, Hazards 27 Conference, Birmingham UK, May 10-12, Institution of Chemical Engineers.

Marsh LLC (2016) *The 100 Largest Losses 1974-2015*, from https://www.marsh.com/us/insights/research/the-100-largest-losses-in-the-hyrdocarbon-industry-1974-2015.html.

McLeod R.W. (2015) *Designing for human reliability: Human Factors Engineering for the oil, gas and process industries*. Gulf Professional Publishing.

McLeod R. (2016) *Issues In Assuring Human Controls in Layers-of-Defences Strategies*, Chemical Engineering Transactions, 48, 925-930.

McLeod, R. W. (2017) *Human Factors in Barrier Management: Hard truths and Challenges*. Process Safety and Environmental Protection. Institution of Chemical Engineers. In Press.

McLeod, R. W., and Bowie, P. (2018) *Bowtie Analysis as a Prospective Risk Assessment Technique in Primary Healthcare. (forthcoming: Policy and Practice in Health and Safety, 2018)*.

NOPSEMA (2012) *Control Measures and Performance Standards* (N-04300-GN-0271), available from the Australian National Offshore Petroleum Safety and Environemntal Management Authority website (www.nopsema.gov.au).

NORSOK (2008) *Technical Safety*. NORSOK Standard S-001.

Perrow, C. (1999) *Normal Accidents*. Princeton University Press.

Pfeffer, J. (1998). *The Human Equation*. Boston, Mass: Harvard Business School Press.

Pitblado, R. (2011) *Global process industry initiatives to reduce major accident hazards*. J Loss Prevention in the Process Industries, 24, p57-62.

Pitblado, R., Potts, T., Fisher, M., & Greenfield, S. (2015) *A Method for Barrier-Based Incident Investigation*. Process Safety Progress (Vol. 34, No. 4), pp. 328-334.

Process Safety Leadership Group. (2009, December 11) Retrieved July 5, 2016, from http://www.hse.gov.uk/comah/buncefield/response.htm

PSLG (2009) *Safety and environmental standards for fuel storage sites*, Process Safety Leadership Group, HSE Books, ISBN 978 0 7176 6386 6, UK.

PSA (2013, 2017) *Principles for Barrier Management in the Petroleum Industry*. Retrieved June 2016, from Norway Petroleum Safety Authority: www.ptil.no, updated with clarifications in a memorandum available on the same site in 2017.

Reason, J. (1990) *Human Error*. Cambridge: Cambridge University Press.

Reason, J. (1997) *Managing the Risks of Organizational Accidents*. Ashgate Publishing.

Reducing error and influencing behaviour (HSG48) HSE Books 1999, ISBN 0 7176 2452 8

Salvi, O., & Debray, B. (2006) *A global view on ARAMIS, a risk assessment methodology for industries in the framework of the SEVESO II directive*. J Haz Mats, 187-199.

SEPA (2016a) Scottish Environment Protection Agency, All Measures Necessary - Environmental Aspects downloadable from https://www.sepa.org.uk/media/219152/d130416_all-measures-necessary-guidance.pdf

SEPA (2016b) Scottish Environment Protection Agency, CDOIF Guideline Environmental Risk Tolerability for COMAH Establishments downloadable from https://www.sepa.org.uk/media/219154/cdoif_guideline__environmental_risk_assessment_v2.pdf

Tyler, B., Crawley, F., & Preston, M. (2008) *HAZOP: Guide to Best Practice, 2nd Edition*. Institution of Chemical Engineers (IChemE).

INDEX